THE ETHICS OF DIAGNOSIS

Philosophy and Medicine

VOLUME 40

The titles published in this series are listed at the end of this volume.

THE ETHICS
OF DIAGNOSIS

Edited by

JOSÉ LUIS PESET

Instituto Arnau de Vilanova
Condejo Superior de Investigaciones Cientificas, Madrid

and

DIEGO GRACIA

Complutensis University of Madrid

KLUWER ACADEMIC PUBLISHERS
DORDRECHT / BOSTON / LONDON

Library of Congress Cataloguing-in-Publication Data

The Ethics of diagnosis / edited by José Luis Peset, Diego Gracia,

    p.   cm. -- (Philosophy and medicine ; v. 40)
  Includes index.
  ISBN 0-7923-1544-8 (HB : alk. paper)
  1. Diagnosis--Moral and ethical aspects.   I. Peset Reig, José
Luis. II. Garcia, Diego. III. Series.
  [DNLM: 1. Diagnosis. 2. Ethics, Medical.   W3 PH609 v. 40 / WB
141 E84]
RC71.3.E8 1992
174'.2--dc20
DNLM/DLC
for Library of Congress   91-35370

ISBN 0-7923 1544-8

Published by Kluwer Academic Publishers,
P.O.Box 17, 3300 AA Dordrecht, The Netherlands.

Kluwer Academic Publishers incorporates
the publishing programmes of
D. Reidel, Martinus Nijhoff, Dr W. Junk and MTP Press.

Sold and distributed in the U.S.A and Canada
by Kluwer Academic Publishers,
101 Philip Drive, Norwell, MA 02061, U.S.A.

In all other countries, sold and distributed
by Kluwer Academic Publishers Group,
P.O. Box 322, 3300 AH Dordrecht, The Netherlands.

Printed on acid-free paper

Printed in the Netherlands

TABLE OF CONTENTS

SECTION IV / COMPUTER AUGMENTED DIAGNOSIS

SECTION V / THE ETHICS OF DIAGNOSIS IN THE POST-MODERN WORLD

PREFACE

This volume is the product of a moveable intellectual feast of reflections on the intertwining of evaluation and explanation in health care. The themes addressed were first encountered by some of the participants in discussions in the late 1960s. In particular, the volume is in debt to conversations between one of the series editors and Corinna Delkeskamp-Hayes in Bonn in 1969 and early 1970. Over the last two decades she contributed in many ways to the discussions that produced this volume. Reflections on the interplay of values, theories, and facts constituted the primary focus of the first volume in this series, *Evaluation and Explanation in the Biomedical Sciences*, published in 1975, in which a number of the contributors to this volume participated. These themes were developed further in subsequent volumes and at various meetings in different parts of the world. One of the crucial meetings in the development of this volume was the Simposio Interdisciplinar Sobre Filosofía y Medicina, organized through the support of the Fundación Juan March (Madrid), the Fritz Thyssen Stiftung (Cologne), and the Instituto Arnau de Vilanova (Madrid) and held at the Fundación Juan March, March 21–23, 1979, with the theme "Etica del diagnostico". Special gratitude in regard to this conference is due Eric Cassell, Valentín Corcés, Dietrich von Engelhardt, Enrique Nájera, Alberto Oriol, Carlos París, Diego Ribes, Ricardo Saiegh, José M.ª Segovia de Arana, and the late Ignacio Ellacuría, S.J. The series editors are in particular debt to Pedro Laín Entralgo, who generously served as President of the conference, and José Luis Peset and Diego Gracia, who served as secretaries.

The conversations and reflections developed at this symposium led to others, and ultimately to the essays in this volume. Along the way, discussions occurred again in Germany, this time at the Institute for Advanced Study in West Berlin, in 1989. They were sustained in Texas with the help of Mary Ann Gardell Cutter. Special thanks are owed as well to the *New England Journal of Medicine* for permission to reprint the material found in the Appendix to Kenneth Schaffner's article. The result is a volume that bears only a limited resemblance to any of the

vii

preceding particular discussions. Yet, the final product is in debt to them all. Most particularly, the series editors are in debt to George Khushf, who orchestrated the final discussions, fashioned the essays into their final form, and made many helpful suggestions regarding the Introduction that follows.

March 2, 1992   H. TRISTRAM ENGELHARDT, JR.
                                                  STUART F. SPICKER

H. TRISTRAM ENGELHARDT, JR.

INTRODUCTION

Knowledge is always value-laden. It is value-laden in the sense that the acquisition of knowledge involves costs. Knowledge is value-laden as well in that the structure given to information derives in part from epistemic and non-epistemic values, from the ways in which one regards certain forms of knowledge as having greater excellence or greater usefulness. Knowledge takes on value from its implications for action. Various values, goals, and purposes cast their light and shadow across that which is known. What we seek to know and how we structure what we see depends on what we expect to see and want to see. In order to move from poorly structured to well-structured problems, one needs already to know what will generally count as information or noise. Background assumptions direct the psychology of discovery, so that one recognizes certain things and ignores others. Choices among ways of seeing the world have costs and are directed by values. Diagnosis is value-laden and invites moral reflection.

Clinicians seek to know not in order to know truly, but in order to act effectively. Physicians are interested in knowing truly, insofar as knowing truly is useful for their therapeutic ends. Which is to say, the medical sciences are applied, not pure. They are embedded in practices and institutions aimed at changing the world. Because these practices are part of large-scale social practices, concerns about costs and benefits become quite salient in health care policy discussions. But even at the very micro-level of a patient and a physician, costs, benefits, goals, values, and social expectations must be considered. Medical knowledge must be evaluated in terms of its likelihood of being true or false and the costs of error. To acquire more information in order to make a more reliable diagnosis is itself not without significant costs. Yet delay itself can be costly. This point is made in Eugene Boisaubin's essay, where he underscores the need of turning first to consider those diagnoses for which a quick intervention is important in being life- or health-saving. In addition, one must bear in mind the morbidity and mortality costs of diagnostic interventions. If to establish a diagnosis one must perform an invasive procedure, then one must consider if making the diagnosis is

1

J. L. Peset and D. Gracia (eds.), The Ethics of Diagnosis, 1–10.
© 1992 *by Kluwer Academic Publishers. Printed in the Netherlands.*

worth the risk of discomfort, pain, and perhaps death involved. Because the process of establishing a diagnosis imposes monetary costs and often morbidity and mortality costs, one must consider how successful the treatment will be if the diagnosis is made, and how bad the disease will be if left untreated. Therefore, as Edmund Pellegrino, Kenneth Schaffner, and Henrik Wulff underscore, one must take into account the context of the patient, when one frames diagnoses.

This volume explores the ethical and value questions associated with medical knowledge, in particular, with diagnosis. The essays bring to the reader the obvious but often under-explored circumstance that important moral, ethical, and value decisions are involved in choosing among different ways of coming to know, of knowing, and of regarding the problems of patients. Bioethics thus includes the bioethics of medical knowledge, an area where moral theory and the theory of knowledge, evaluation and explanation intertwine. To begin with, there are moral issues in knowing truly in the sense of appropriately minimizing errors of diagnosis. Concerns with knowing truly in the sense of making diagnoses on the basis of sufficient information and careful analysis are, as Lain Entralgo shows, at least as old as Greek medicine.

Such concerns compass not just the due diligence that physicians should have in making diagnoses, but issues in the sociology of medical knowledge in the sense of the social conventions accepted by physicians in establishing particular diagnoses. Thus, for example, to see or regard a particular pathological specimen as benign or malignant, one must already have decided how many mitotic figures with what amount of aneuploidy per high power microscopic field justify the conclusion that there is a malignancy. The criterion axis employed in distinguishing between malignant and non-malignant findings depends on judgements made against background views regarding the appropriate balancings of the costs of over- versus under-treating for possible malignancies. This therapeutically-oriented way in which medical knowledge and reality is organized is especially clear in the staging and grading of cancer. In grading cancers, one divides a continuum of cell abnormalities into units to aid physicians in choosing therapies and providing prognoses. Similarly, one stages cancers into neat categories in order better to pursue prognostic and therapeutic goals. There are also choices to be made among alternative, theoretically-based ways of construing medical diagnosis. Many diseases can be seen as infectious, genetic, or environmental diseases. Depending on how one characterizes particular diseases (e.g., as infectious, genetic, or environmental), one predestines patients to the care of particular medical specialties. The way one regards medical findings is tied to what one wishes to tell patients and to do with patients. Decisions about how to know become fraught with moral implications. One is not simply knowing a fact but choosing among different ways of construing facts, which have different costs and benefits for patients, society, and physicians.

How one categorizes or diagnoses patients influences how patients regard themselves and are regarded by others. For example, to be diagnosed a diabetic is in part to be embedded within a set of dietary and clinical expectations. What were taken-for-granted, everyday happenings become reevaluated in terms of special theoretical expectations. Indeed, in chronic diseases most particularly, patients are inducted by a diagnosis into a set of obligations, expectations, and explanations. Also, the way in which a diagnosis is conveyed to a patient may lead patients to being seen primarily as cases of particular diseases, thus isolating them from the full experience of themselves and their illness. Such patients can experience a dehumanizing attention focused on the derangement of a particular organ or organ system to the exclusion of their illness. Medical diagnoses and the act of conveying a diagnosis, as Mainetti, Leder, Spicker, and Bole underscore, determine the ways in which illness and disease are experienced, both by the patient and those around the patient.

In all of this, as Bole recognizes, there are choices to be made among the goals to be achieved by medical knowledge. These choices determine how physicians approach the problem of diagnosis, because diagnoses are always tied to therapeutic expectations. There are choices to be made among the ways one can structure medical knowledge. There are choices to be made among different diagnostic frameworks and ways in which theories can organize and direct the diagnostician. Like the medical student learning to "see" the structure of cells in histology, physicians and patients learn to see everyday happenings as well as laboratory findings in terms of diagnostic and therapeutic expectations and interests.

In summary, how one frames diagnostic expectations will also have an impact on how one understands and approaches particular groups of patients. This will in part be the case because of the psychology of discovery, because one sees what one expects to see and neglects what one does not anticipate. This will also be the case in terms of the values one gives to different circumstances in different social institutions. Thus, one finds as one would perhaps expect that diagnostic taxons emerge, supported by clinical and laboratory findings to support diseases that we can now recognize to have been made because of the social expectations of the time. One might think, for example, of the disease of masturbation [2]. One might think, as well, of the special diseases framed for slaves in the South before the Late Unpleasantness [1]. An interesting feature about Cartwright's proposals for the diseases of slaves is that one does not know whether they are being employed as a form of repression similar to the diagnosis of sluggish schizophrenia in the Soviet Union [4], or whether in fact they were a clandestine way of protecting slaves against punitive actions by their masters through placing them in a sick role, thus excusing them from infractions that would otherwise elicit punishment, not treatment. Finally and importantly, one must underscore the ways in which historically and culturally embedded understandings influence the ways in which men and women are differentially treated and their diseases regarded. If one looks to the 19th century, it is easy to identify the "non-objective" intrusion of cultural idiosyncrasies into the ways in which women were regarded medically and their illnesses understood [6]. This intrusion of socially supported expectations opens up a wide range for contemporary exploration regarding the ways in which diagnoses are framed regarding women and the ways in which studies are undertaken to understand diseases. An understanding of the social construction of medical knowledge opens doors to social critique and reevaluation in terms of individual and social goals and expectations. The third section explores the foundational elements of these concerns.

The concept of disease itself is also value-dependent, for to talk about successful human adaptation, one must not only specify the environment within which humans are adapted, but the goals of adaptation. The latter will not be dependent simply on such biological "goals" as maximizing inclusive fitness, but on accepted, usually implicit ideals of human function, freedom from pain, human grace of form and action, and life expectancy. The lines between what will count as disease and normal variation are determined by background views of what levels of function and characteristics of form are to be expected for individuals of a particular age and sex, what freedom from pain should characterize human existence, and what life expectancy is appropriate. In this volume, Thomas Bole addresses special attention to this cluster of issues.

Diagnosis creates social reality. Diagnosis is a performative act, much like the act of a sheriff arresting a suspect. The person diagnosed is placed within a set of social roles and expectations. Though the patient may be blamed for having become sick, the sick role excuses its members from being sick, as long as steps are taken to seek treatment from socially recognized therapists. A diagnosis places an individual in a particular social space, framed by particular rights, duties, and expectations [5]. The decision to make a diagnosis is a social decision with the moral concerns that attend it. One must in particular note that diagnoses are rarely made with perfect certainty, so that one must consider with what level of certainty one must know before one may make a diagnosis of a particular sort, given its social and therapeutic consequences. Different diagnoses entail different therapeutic warrants with different morbidity, mortality, and social risks.

The first section of the volume addresses the history of concerns about framing diagnosis. Lain Entralgo's and Gracia's papers review the quite different approaches of ancient Greek and Christian medicine to the problem of medical knowledge. Greek physicians saw as their responsibility the competent disclosure of what could be known truly. There was an obligation to and regarding knowledge. The Greek physician was required as well to recognize the limits of the craft and refrain from interventions that could not provide a benefit. In all of this, the Hippocratic physician was directed by love of man and love of the art, by the goal of helping the patient and the goal of perfecting the art of medicine. In these twin goals, there was already the beginning of the tension between the physician as a clinician and the physician as a scientist.

Unlike the secularized approach to the world, which directed the concerns of the Hippocratic physician, illness was understood in ancient Israel to be a consequence of sin. In particular, differential diagnosis and the treatment of illness was in great measure shaped by concerns with ritual purity and the interventions of the priest. This view was transformed by the special Christian accent on the salvific and healing work of God, so that in the 2nd century, when Hellenic and Christian interests met, there was a combination of the Christian ethic of agape, the Hippocratic concern with philanthropia and the Greek naturalistic ethic with its accent on physis, the norms of nature. In the Middle Ages, this mixture of concerns led to an interest in medical diagnosis and technical competence that was to be in conformity with good order or with a good judgement of the physis of man and the community. Nature was the criterion of morality. There was as well an interest in a special ethic of cure and care for the poor and sick. These two clusters of concerns were tied to social differences, the first being directed primarily to the upper strata of society and the second to the lower.

Contemporary medicine, as Albarracín and Peset show, developed out of important social and scientific changes. These developments produced distinctions between civic and religious duties, such that medicine came to be regarded as a science independent of religion, directed by its own codes of medical ethics or deontology. Medical knowledge became secular knowledge, in contrast with the view that saw right knowledge to be derived from an exegesis of scripture in addition to observation of the world. These secularizing developments in medical knowledge were tied to industrialization, such that concerns with diagnosis and treatment became intertwined with concerns regarding medical costs to governments, societies, and companies. Social security systems were enacted, especially in Europe, in order to meet demands by workers, creating formalized diagnoses and warranted therapies within explicit reimbursement systems. As Amundsen underscores, standards of diagnoses are justified in terms of nosological systems. But as nosological systems became incorporated within social insurance systems, they took on a special social character open to bureaucratic adjudication, democratic debate, and economic distortions. One might think of reimbursement systems such as the diagnosis-related groups of American Medicare, where the choice among diagnoses determines the reimbursement to hospitals. The choice between the diagnosis of angina and unstable angina is a choice between quite different levels of reimbursement.

The second section explores the ways in which diagnostic expectations frame the experience of illness and disease by physicians, patients, and societies. As Mainetti shows, diagnoses present one's body and capacities in terms of negativities, in terms of shortcomings. The choice of diagnostic languages is a choice among ways of experiencing embodiment, among phenomenologies of illness and disease. As Leder notes, diagnosis is an ontological event, which ratifies a particular account of reality. Diagnoses tie the present with the future in terms of the prognoses they warrant. As has already been underscored, diagnoses also, in their contemporary, highly technological character, run the risk of dehumanizing individuals in focusing on particular organs or organ system failures, while ignoring the whole person and the full experience of illness. Diagnosis can in such circumstances truncate or restrict attention to only a portion of a patient's illness by providing legitimation for only a portion of a patient's experience of suffering. In particular, physiological or anatomical explanations can legitimate or delegitimate particular complaints. The result can be a ranking of complaints and patients in terms of which complaints are real or serious and which are trivial. The triumph of the basic medical sciences in the 19th century, which expanded the prognostic and therapeutic capacities of medicine, at the same time supported a reevaluation of the worried well and the vexed ill, for whose complaints one often cannot provide anatomical and physiological groundings. The ideology of laboratory diagnosis cast its shadow on the treatment of patients.

Spicker and Bole develop further the points introduced by Mainetti and Leder concerning the role of values in the experience of the body and bodily norms. Spicker distinguishes between the process of diagnosis and the act of diagnosis. The latter places a patient within a normative and social category and can thus transform the life of the patient. In addition to such internal norms of medical diagnosis as the obligation (1) to overcome observer bias, (2) to tell the truth, (3) to contribute to the medical literature, (4) to avoid inordinate gain from the process of diagnosis (e.g., through the use of physician-owned diagnostic laboratories), and (5) to develop diagnoses with due diligence, the physician must attend to the consequences of diagnosis for the self-image and integrity of patients. Bole directs special attention to the appropriate ways in which patients should be involved in choices about the values that frame their diagnoses.

The third section explores the interplay of evaluation and explanation in terms of social expectations and structures, whose interplay lies behind the social construction of medical knowledge. Wartofsky gives special attention to the interaction among factual descriptions, normative considerations, and social structures within the theoretical undertakings of medicine. His essay demonstrates that elements of this interplay cannot be understood in isolation. For example, medicine contributes to the social definition of person, while the social definition of person structures medicine. The interdependence between medical definitions and social concerns is further explored by Sass with reference to the development of definitions of death and concepts of disease. As Sass suggests, the introduction of computer diagnoses helps to disclose the culture-dependent character of medical concepts because it becomes ever more plain that nosological taxa represent instrumental, not natural categories. This set of studies is completed by Mary Ann Cutter, in an examination of the role of cost considerations in interpreting PAP smears. This examination displays the role of non-epistemic considerations in the establishment of indications for medical intervention.

The penultimate section takes the themes from the rest of the volume and brings them to a focus around the use of computers in diagnosis. As Pellegrino notes, the intrusion of a computer in the diadic relation of physician and patient need not displace humanistic values or radically alter the character of medicine. Instead, computer-assisted diagnosis can offer a special, technological form of medical consultation, while highlighting the contribution of the physician in determining what is the right and good action for a particular patient under care. As Pellegrino shows, final choices among possible diagnoses will always have to be made by physicians in light of the benefits and dangers associated with different diagnostic alternatives. The technological contributions of the computer can help prevent physicians from becoming mere technologists because what the computer offers still requires the interpretive and evaluative skills that a physician must contribute.

Kenneth Schaffner further delineates the role of judgement and clinical acumen in computer-assisted diagnosis by exploring particular computer programs and the ways in which they involve choices among different ways of construing knowledge and moving toward medical decision. These choices reveal the constructed character of medical knowledge, and the extent to which computer diagnostic programs reflect choices among theoretical constructions and goals that often lie hidden in everyday practice. The use of computer programs requires a disciplined assessment of the goals of diagnoses because, as Wulff stresses, the computer encourages the clinician to move from uncontrolled biological or empirical thinking to controlled or disciplined empirical thinking.

Clinical decision theory, whether or not employed via a computer, can underscore the humanistic goals of medicine, if the physician recognizes the central role that patient' values play in determining costs and benefits. Indeed, as Murphy shows in his examination of the contectual and theory-determined character of all medical findings, the nature of disease and the objectives of diagnosis are disturbingly vague. The use of computer-assisted diagnosis, as a consequence, reveals how much must be done to understand better both the clinical and mathematical aspects of diagnosis. Finally, Boisaubin underscores the ways in which computer programs must incorporate the goal-directed character of medical knowledge. For example, computer- or decision-theory-assisted diagnosis ought to move clinicians towards first making diagnoses of serious illness for which there is a good treatment and only then encourage the exploration of diagnoses of untreatable or minor diseases.

The volume closes with a philosophical reflection on post-modernity and its implications for medical diagnosis. George Khushf reminds the reader that the plurality of metaphysical and moral visions, which define the contemporary predicament, have dramatic implications for the ethics of framing diagnoses. Insofar as diagnoses depend on different social and value expectations, one may have an obligation to disclose to patients the social and value biases intrinsic to particular views of medical objectivity. If, to paraphrase Lyotard, there is no universal concrete moral narrative, there is also no universal evaluational narrative to direct health care [3]. Khushf proffers a regulative guide for the ethics of diagnosis in a post-modern world. The guide he offers is the practice of attempting intersubjectively to resolve controversies within somatic medicine. In framing this suggestion he borrows inspiration from Kant, recognizing that the guidance must be regulative, not constitutive. Khushf suggests that one can find an intersubjective focus around which to integrate the divergent metaphysical and value commitments, which direct patients in the diverse moral communities of a post-modern world. Though any understanding of somatic medicine will always be socially and historically conditioned, still somatic medicine offers, so Khushf argues, a focal point around which to coordinate the energies of physicians and patients who bring with them differing understandings of the meaning of reality and the significance of human life.

The volume thus concludes with an important acknowledgement: one cannot discover a hard, factual core that is sufficient to guide health care policy or the ethics of diagnosis. One must always make reference to particular content given in particular interactions or to the special metaphysical, moral, and social commitments that physicians, patients, and communities bring with them. There is an ethics of diagnosis because there are important choices to be made regarding how to know. Knowing is never an unambiguous act. It is always freighted by moral decisions and commitments. This is never more true than in health care, where there is an intertwining of concerns about knowing truly, acting effectively with the permission of those involved, and acting so as to achieve important goods in the face of human limitation, suffering, and death.

August 1991

BIBLIOGRAPHY

1. Cartwright, S.A.: 1851, 'Report on the Diseases and Physical Peculiarities of the Negro Race', *New Orleans Medical and Surgical Journal* **7** (May), 691–715.
2. Engelhardt, Jr., H.T.: 1974, 'The Disease of Masturbation: Values and the Concept of Disease', *Bulletin of the History of Medicine* **48** (Summer), 234–48.
3. Lyotard, J-F: 1979, *La Condition postmoderne: rapport sur le savoir*, Les Editions de Minuit, Paris.
4. Mersky, H.: 1990 '"Sluggish Schizophrenia" and Outmoded Diagnosis', *London Psychiatric Hospital Bulletin* **5**, 12–19.
5. Parsons, T.: 1958, 'Definitions of Health and Illness in the Light of American Values and Social Structure', in E.G. Jaco (ed.), *Patients, Physicians and Illness*, Free Press, Glencoe, Illinois, pp. 165–187.
6. Smith-Rosenberg, C., and Rosenberg C.: 1973, 'The Female Animal: Medical and Biological Views of Woman and her Role in Nineteenth-Century America', *Journal of American History* **60** (Sept.), 332–56.

SECTION I

HISTORICAL PERSPECTIVES

PEDRO LAÍN-ENTRALGO

THE ETHICS OF DIAGNOSIS IN ANCIENT
GREEK MEDICINE*

I

Every human act that implies a relation with the external world, as is the case with medical diagnosis, presupposes a cognitive distancing of the agent from the area of reality to which the act refers. Nobody, for example, could set out to travel to a city without a certain prior idea of what that city is, even when such an idea may be no more than an erroneous or imprecise conjecture. Not even the man who sets out exploring with the maxim, "Well, let's see what happens", is exempted from this rule. To the question implied in that distancing, there are two, qualitatively distinctive answers: one empirical, constituted by what the corresponding experience of the world has taught; and the other interpretative, formed by what is for that agent the reality of the particular experience. In the interpretative moment, up to four types of ideas can be distinguished: the purely empirical or the "reduplicatively empirical", one could call it – of those who with greater of lesser deliberation do not want or do not know how to adhere to anything but the data afforded by sensorial experience; the magical, of those who in some manner or other have recourse to the intervention of preternatural and superhuman potencies in their judgements; the imaginative, of those who appeal to notions which, while not possessing a preternatural character, cannot be rationally justified; and finally, the rational, of those who only by means of experience and reason set out to interpret what they perceive. In the actual conduct of those who interpret and act, an interplay of these four types of ideals of interpretation in one form or other is usually found.

Understanding the word "diagnosis" in its broadest sense enables one to appreciate the roles played by the four conceptual ideas. The history of medical diagnosis confirms their presence. Two distinct periods in this history should be distinguished: that prior to the Alcmeonic-Hippocratic medicine, and that which begins with this medicine. In the former, whose duration extends from the origin of the human species to Greece of the sixth and fifth centuries B.C., the cognitive distancing of the healer from the reality of the ill person is filled by something that does not deserve

13

J. L. Peset and D. Gracia (eds.), The Ethics of Diagnosis, 13–18.
© 1992 *by Kluwer Academic Publishers. Printed in the Netherlands.*

the name "diagnosis" in the strict sense of the term; it is no more than a quasi-diagnostic nomination of the state of the patient, which corresponds to an empirical, magical, or imaginative interpretation of the disease or illness state. No matter how subtle the ancient Egyptian, Chinese, or Indian healer was in his approach to the ill person, he was unable to surpass this pretechnical mode of understanding disease. In contrast, the medicine initiated by Alcmaeon of Crotona and the Hippocratics invokes the verb *diagignoskein* and the substantive *diagnosis* as technical terms to designate the medical knowledge of "being ill". To be sure, technicians or healers included at times in their interpretations elements of a purely imaginative or *lato sensu* magical character. These are exemplified by the contents of many Cnidian diagnoses, the procedure of the Byzantine Alexander of Tralles (despite the soberness or restraint of his clinical technique), and the *ens deale* of the modern Paracelsus. Nevertheless, interpretation in terms of experience and reason can be seen to dominate the diagnostic approach of the healer. This approach, captured in diverse socio-historical situations in ancient Greece, Ancient Rome, Byzantium, the Islamic world, European Middle Ages, and the modern world, continues to be the approach used in the diagnostic technique of the modern physician.

In light of this partition in the history of medical diagnosis, consider some of the problems regarding the ethics of diagnosis. Every human act, simply by being human, possesses an ethical dimension. Since the act of diagnosing, as its name indicates, is preponderantly cognitive, its condition of "moral goodness" and "moral evil" is understood in terms of an "ethically correct response" and an "ethically culpable error", respectively. A diagnotic judgment may be technically good, if it correctly expressed what is taking place in the reality of the ill person, and ethically bad, if it has been obtained by means that are coercive from the moral point of view; conversely, a diagnostic judgment may be technically bad and ethically good (or at least not culpable) when the physician for whatever reason, so long as there is no question of negligence or evil intent, has erred in his attempt to establish that diagnosis.

At this point, three important questions arise. First, suppose the physician commits a culpable diagnostic error; before whom or before what is the physician to be held culpable? Second, how are the culpable errors committed by the physician realized in the field of diagnosis? Third, how does the fault inherent in such errors come to be established and sanctioned?

In the history of medical ethics, the various modes of response to these queries have been taken up under the rubric of the ethics of diagnosis. The rest of this paper considers how the medicine of classical Greece responded to these queries.

II

The *teknites* of Hellenic medicine – professionals devoted to the practice of *tekhne iatrike* – considered themselves morally obliged in all medical acts towards *physis*, or that which constitutes divinity, insofar as this is a constitutive reality endowed with internal "reason" or *logos*, according to the clear-cut teaching of Heraclitus. On this view, medical diagnosis was held to constitute the proper knowledge of the disorder produced by the disease in the *physis* of the patient – the *logos* proper to that *physis*. For that reason, the author of *On Places* argues that the *physis* is the principle of the *logos* of the physician.

Numerous writings in the Hippocratic collection refer to the traditional gods with respect, and even encourage swearing by them. However this residual devotion towards the gods of Olympus, so often acknowledged by the Greek authors after the fifth century B.C.,[1] is usually depicted as a conventional or tacit agreement between critical religiosity – the religiosity of Plato, Aristotle, and other philosophers for whom "the divine" is the *physis* – and the traditional or popular religiosity to which these gods of Olympus belong. Following the profound intellectual and religious transformation that occurred in the Hellenic mind in the sixth and fifth centuries B.C., the critical Greek recognizes the supreme tribunal over questions regarding moral responsibility to be in the superhuman, an unappealable "reason" (*logos*) of the universal *physis*: "reason" is made dynamically manifest in the process of the *kosmos*, and its transgression under the form of *hybris* can be physically sanctioned. As a disputed text of *Prognosis* reads: "The nature of diseases must be known, and the extent to which they surpass the strength of the body, and at the same time if there is anything of the divine, *ti theion*, in them" (*Prognosis*, Littre II, 112). It is not accidental that the oldest Latin versions of this passage translate the Greek *theion* by *ratio mundi*, or "supreme reason of the cosmos". Whether in fulfilling what this "reason" prescribes and what observation enables one to know, or in bravely proceeding on the always uncertain frontier between chance (*tykhe*), invincible fatality (*ananke*), and culpable transgression of that limit, the conscience of the Hippocratic physician is placed before the tribunal of the "supreme reason of the cosmos" to establish the moral quality, "good" or "evil", of his diagnostic and therapeutic conduct. This is what one is led to conclude, given the ambivalent semantics of the terms *hamartia* and *hamartema*, which designate both error and sin, and in any given context emphasize one or the other.

III

Failure to perceive the clinical signs of an incurable condition (*kat' ananken*), which is fatally determined by the divine *physis* and which is manifested in the state in which the patient finds himself, can be a culpable negligence. According to the text *On the Art*, the *tekhne* of the physician requires, among other things, that he "abstain from treating those who are dominated by the disease, because in these cases art can do nothing " (Littre, VI, 2). "It is necessary not to think, say, or do unfeasible things", adds the author of *On Disease* (Littre VI, 140). And since the sovereign and unappealable designs of the *physis* are evident to the observer through certain signs, the physician is technically – and morally – obligated to discern and take account of them in his exploration of the ill person's body.

Carelessness in the search for clinical signs in any morbose affection, whether caused by invincible fatality (*kat' ananken*) or by surmountable chance (*kat' tykhen*), offends against the physician's aspiration to have his practice ruled by the two basic principles of Hippocratic medicine: "love of man" (*philantropia*) and "love of art" (*philotekhnia*). The minute detail and reiteration of the rules regarding what the clinical examination and the anamnesis should consider shows the character of technical duty, and hence ethical duty, which these rules had for the Greek physicians. At least, this is the case for Plato with respect to the physician's practice with free and well-off citizens.

The pursuit of fees without an interest in the clinical care in question and its contribution to medical knowledge is a culpable error. "Let us not think about salary but about the desire to seek instruction" (*Precepts*, Littre IX, 258). In other words, if the physician does not attempt to help the patient – Recall the famous imperative to "help, or at least do no harm" (*Epidemics*, Littre II, 634–636) – and to perfect his knowledge with his diagnosis, his conduct will be neither technically nor ethically proper.

IV

As previously discussed, moral responsibility presupposes the existence of an ultimate "who" or "what" before which this responsibility is judged. In a number of archaic cultures, this ultimate "who" or "what" is the gods, the powers that direct the world in a supreme way. These powers are represented by clearly defined men or castes (e.g., kings, priests, etc.), to whom the knowledge of the sanction of the punishable actions falls. This is evident in Assyrian medicine's Hammurabi Code. In contrast, classical Greek medicine does not offer any appropriate analogies. The "error-sins" the physician might commit in his practice either against the cosmic order of the *physis* (the *kosmos*) or against its political order (the *polis*) were not punished by those who exercised power. In other words, classical Greece was without a tribunal to judge the physician 's professional responsibility. Does this mean that the *hamartemata* lacked all sanctions? No, but the sanction of the error-sin or *hamartema* of the Greek physician took on a subjective and a social form.

The physician who committed any transgressions against the proper diagnostic procedure should feel in it – if in truth he were responsible and knew the didactic texts of his profession – a sharp feeling of guilt. "If the physician does not treat the patient well or if he does not know his illness, and the ill person is overcome by it, the physician is to blame" (*On Affections*, Littre VI, 220).

In a culture like the ancient Greek where, from Homeric times, so much social importance was given to points of honor and reputation – points reflecting what Dodds (1956) [1] calls a "same-culture" – it was natural to see in the relationship between prestige and discredit a gratifying or punitive sanction of the conduct of the physician, both in the therapeutic order (good or bad results of the proposed cures) and in the diagnostic order (adequacy of knowledge of the illness, and success or error in the prognostic judgment). The Hippocratic Aesclepian, the author of *On Decency*, spoke for all: the wisdom of the physician is valuable when art is carried out "for decorum and good fame" (*On Deceny*, Littre, IX, 226). It is from this ethico-social point of view, as Edelstein (1967) [3] indicates, that the frequent Hippocratic allusions to the physician's fame in the world of the *polis* are to be understood.

In short, diagnosis in ancient Greece had a prominent moral component.

University of Madrid
Madrid, Spain

PEDRO LAÍN-ENTRALGO

NOTES

* The editors wish to express their thanks to Michael C. White for the translation of this essay.

[1] For instance, the so long protracted devotion to Asclepios.

BIBLIOGRAPHY

1. Dodds, E.R.: 1956, *The Greeks and the Irrational*, University of California Press, Berkeley and Los Angeles.
2. Edelstein, L.: 1931, *Peri aeron und die Sammlung der hippokratischen Schriften, Weidmannsche Buchhandlung*, Berlin.
3. Edelstein, L.: 1967, *Ancient Medicine*, Johns Hopkins Press, Baltimore.

DIEGO GRACIA

THE ETHICS OF DIAGNOSIS IN EARLY
CHRISTIANITY AND THE MIDDLE AGES*

I. CHRISTIANITY AND THE ETHICS OF MEDICAL DIAGNOSIS

The approach to illness for the ancient people of Israel, as for all archaic
Semitic peoples, is of an ethico-religious nature since these peoples view
illness, as the consequence of sin. Hence the search for the cause of
illness, the "etiological diagnosis", is carried out by means of a systematic inquiry into the moral precept that has been transgressed. This is the
procedure followed in the medicine of Mesopotamia as well as that of
the people of Israel.

The legal canon of the Israelites is to be found in the Pentateuch, the
Torah or Book of Law, and the most highly concentrated expression of
this canon figures in the Decalogue or Law of Moses. However, the
Torah conserves not only one but two versions of the decalogue, the one
"Deuteronomist" (Deut. 5: 6–21) (which forms part of what from
Wellhausen onwards has been known as "source D", probably set down
in the seventh century B.C. by the Levites of Galilee) and the other "sacerdotal" (Exod. 20:2–17) (belonging to the Priestercodex or "source P",
written down toward the sixth or fifth centuries B.C by the priests of
Jerusalem). The views proper to these two sources are quite distinct. The
former is marked by a "religious" and "prophetic" character, while the
latter is more "ritualistic" and "juridical". Thus, the ethical problem of
illness in general, and the ethics of diagnosis in particular, is typified in
two disctinct ways in the Old Testament tradition.

In the language of Deuteronomy, health appears as a gift from Yahweh
and illness as a debt owed to him on account of the transgression of the
law imposed on his people. Corporal or somatic illness is the manifestation of a sin of the heart: such is the "etiological diagnosis" (Deut.
28:15). Besides this there exists a "differential diagnosis", which leads to
a differentiation between the different types of illness (Deut. 28: 21–22).
Naturally, there is also an etiological diagnosis of the state of health,
which in turn is seen as due to the fulfillment of the law of Yahweh
(Duet. 7:12–15).

Following the Babylonian exile, little by little the Israelite religiosity

19

J. L. Peset and D. Gracia (eds.), The Ethics of Diagnosis, 19–27.
© 1992 *by Kluwer Academic Publishers. Printed in the Netherlands.*

begins to lose that lively and creative character proper to the prophetic
period; it becomes scholastic, legalistic, and juridical. The place of
primacy passes from the internal intentions to formal ritual. It is the language of Leviticus, composed by the sacerdotal caste of Jerusalem, who
are more concerned with the rites of the temple than with true religiosity
of the heart. The ancient language, that understood health as a gift from
God and illness as a debt, is substituted by another where health is interpreted in terms of purity and illness is seen in terms of stain or impurity,
as a sign of corporal decay, as a sign that death is threatening. Yahweh is
life, while death and illness are signs of possession by Satan. As against
the servant of God, the pure one, there is the slave of Satan, the impure
one. Because he is impure, the ill person is segregated from the cultural
community of Israel. As against the religious language of "gift-debt",
there now arises the juridical language of "pure-stained". If in the former,
the diagnosis was fundamentally "etiological", since it tried to establish
clearly that the cause of health was the fulfilment of the law of Yahweh,
now the "differential diagnosis" takes the primary place for which
Chapters 13 to 15 of Leviticus provide evidence. It was this language of
purity-impurity that was prevalent at the same time of Jesus, as can be
seen from the laws of Qumran.

Jesus rejects all the Levitic language of the pure-stained (Mark 7:1–16;
Matt. 15:1–20). It might be thought that this rejection is made in favor of
the language of Deuteronomy, in which case illness would continue to be
interpreted as a debt or sin and health as a gift. This however, is not the
case. In the Gospel of St. John, we have the scene of the man blind from
birth where the disciples of Jesus question their master: "Rabbi, who hath
sinned, this man or his parents, that he should be born blind?" Jesus
replies: "Neither hath this man sinned, nor his parents; but that the works
of God should be made manifest in him" (John 9:1–3). Despite the relatively late date in which the Gospel of St. John was completed, current
textual criticism unanimously admits the antiquity of these verses.[1]
According to Boismard, verse 1 belongs to document C (composed about
the year 50 A.D.), while verses 2 and 3 belong to the gospel called John
II-A (set down between the years 60 and 65 (A.D.) ([2], pp. 67, 68,
246–262). The message that these verses transmit is not therefore a personal and late opinion, but one of the nuclear elements of the tradition of
Jesus, which can be seen to have partial parallels in the other gospels
(Luke 15:1–5; Mark 2:1–12; Luke 11:20; Matt, 12:28) [5]. The *kerygma*
of Jesus neither directly nor indirectly considers illness as a consequence

of a moral trangression or a sin. The ill person is not looked upon as an impure being that must be excluded from the cultural and social community, but as a privileged place or object for the manifestation of the salvific work of God. This work is Jesus, the Saviour whose practice overrules and surpasses the Old Testament practices. And this new practice is essentially practice with outsiders, the poor, and the ill (Matt. 11:45). To deal with the sick is an essential feature of Messianic practice; consequently, the sick attain a privileged place in Christian religiosity (Matt. 4:23). The ill person is visited (Matt. 25:367), attended, and helped. The "manifestation of the work of God" is, therefore, the manifestation of *agape*, or *caritas*, of benevolence and beneficence. It is the new commandment (John 13:34–35). This commandment sets up a norm of life and, therefore, a norm of ethics, but not an ethics of a "deontological" type, as was the case of the Old Testament law (Matt. 5:17), but of a "teleological" type [22]. The great ethical novelty brought about by the New Testament is the sublation – which is both an annulment and a fulfillment: "Charity is the law in its fullness" (Rom. 13:10) – of the ancient "deontological ethics" in a "teleological or preferential ethic" (Rom. 13:8).

This, indeed, typifies the problem of medieval diagnosis and its ethical dimension in a new manner. It is important to note that remains of the Levitic "differential diagnosis" are to be found in the Gospels. It appears that whenever an illness is named in them (e.g., paralysis, leprosy, blindness, bleeding, etc.), it is done in accordance with the diagnostic tradition of the Jewish sacerdotal caste, so much so that the cure has to be checked by the priests of the temple (Matt. 8:4). Nevertheless, the innovation is radical in what has been termed "etiological diagnosis". It is true that, as in the Old Testament, the diagnosis still has a formally ethico-religious meaning, but this now has a distinctively new character. Illness is no longer the consequence of a moral and religious transgression but of a situation of existence dominated by weakness (*astheneia*), one in which, as a consequence, the power of God (2 Cor. 4:7–12) can be more clearly manifested. Illness is par excellence a Messianic place or situation. Hence we see the preoccupation of the early Christian communities directed to understanding illness as a salvific place, a place to act with charity, and, consequently, a place where God, who is love, is present. It is, as in the case of the Old Testament, an "etiological diagnosis" of an ethico-religious type, but of completely new characteristics: the knowledge of illness as par excellence a salvific place. Perhaps also, because of this reason, the technical aspect proper of medical care was neglected, so much so, that in the early Christian communities we may speak of care of the ill person but not of medical care ([10], pp. 355–387).

II. THE ETHICS OF MEDICAL DIAGNOSIS IN THE MIDDLE AGES

The life of the early Christian communities was guided by the sole preoccupation of following Jesus in his life and in his practice, without explicitly raising the question of the relationship of the Christian message with the Greek *logos*. In the second century, a meeting between Hellenism and Christianity takes place, and the people who have been educated in the great Greek culture join the Christian religion, initiating a process whereby Greek categories are assimilated to the Christian message. Theological science, understood as the application of the Greek philosophical *logos* to the analysis of the Christian *Theos*, thus arises. Besides philosophy, the assimilation process extends to literature, science, and, what here particularly interests us, the medicine of the Hippocratic and post-Hippocratic schools. In keeping with the precept of charity that bids the Christian to love and do all the good possible to the person in need, the second-century Christians begin to make use of the Greek *techne atrike* to aid the ill. Such a process of assumption was, however, beset with problems [1]. How could the ethics of charity be articulated with Greek naturalistic ethics? Among the various possible solutions, there was one that was to be adopted. *Agape* or charity, i.e., the ethics of the New Testament, became the "ultimate norm" of mortality, while *physis* (nature), the foundation of Greek ethics, since *dikaiosyne* (justice) is understood as the "natural" ordering of the elements within the *physis*, became the "proximate norm" of mortality. This *physis* embraces not only inanimate or cosmic nature, but also political nature – the order of the *polis* and of the *kosmopolis* as a natural order – and human nature. Thus, Christianity finds itself impelled to accept the Graeco-Roman social structure as natural and desired by God, the Author of the order of nature and also of medical science that studies the nature of man. This accepts, on the one hand, the assumption of the Graeco-Roman medical practice, which clearly differentiated between the wealthy free and the poor serf, and provided technical care only to the former. On the other hand, it also assumes the concrete ethical statute or statutes found in Hippocratic medicine. These were fundamentally two. The first is the "ethic of social prestige" of the Hippocratic text *On the*

Physician, which was to inform the relations of the technical physicians with their patients of the upper social echelons throughout the Middle Ages. The other is that of the "ethic of the good of the patient", of the Hippocratic *Oath*, which was of greater religious character, and which, colored with the precept of Christian charity, was to become the ethical norm of care towards the poor and needy [4]. The former is a "physiological ethic", the latter a "religious ethic". Let us now see how each develops during the Middle Ages.

A. The Physiological Ethics of Medical Diagnosis in the Middle Ages

The technical or scientific medicine of the Middle Ages is the result of the reception, assimilation, and adaptation of the Greek Hippocratic-Galenic medicine by the great medieval cultures, namely Byzantine, Jewish, Arab, and Christian. In all cultures it showed similar characteristics, which enable us to deal with them together as a unit. In all cultures technical medicine extended – if not exclusively, certainly at least in a preponderant way – only to the upper sector of the population, that is to say, to the great civil and ecclesiastic personages and to their respective courts and communities.

The diagnosis carried out by these medieval technical physicians with those patients is practically identical to the Hippocratic-Galenic diagnosis. Thus, a distinction was made between a diagnosis of a preventive or hygienic character and one of a curative character (as a diagnosis of the illness). Let us consider these in closer detail.

(1) In the first place, there is a preventive or hygienic diagnosis. It is a question of getting to know (*gignoskein*) by means of (*dia*) the study of the physis (or nature of the patient) the regime, or way of life, which a person or community is to follow so as not to alter the balance of his nature and, consequently, so as not to contract illness. This hygienic preoccupation, in the case of the more favored social strata of medieval society, gave rise to two very prolific literary genres, one directed fundamentally to civil personages, kings, and feudal lords, namely, the literary genre of the *regimina sanitates*, to which belong, among others, the *Regimen sanitatis ad regem anglorum* of the school of Salerno, the *Regimen sanitatis ad inclytum regem Aragonem* of Arnau de Vilanova, etc. The other literary genre is directed to the *oratores*, and consists of monastic rules or *regulae vitae*, the most influential of which is that of St. Benedict. These rules are in the manner of *regimina sanitates "au*

divine". The ethic at the root both of the physiological diagnoses, from which stem both the *regimina*, and of *regulae* is that of "good order" (the isonomia of Alcmaeon of Croton) or "good adjustment" (*dykaiosyne*) of the *physis* of man or the community. That order and that adjustment are physiological and, therefore, "good", since they are taken from nature as the criterion of morality.

(2) Besides this preventive diagnosis, there is another more usual one, namely, the diagnosis if illness. It deals with getting to know the pathological state by means of the signs and symptoms that the ill person presents. This diagnosis has a technique and a method. But it also has an ethic that is represented in a paradigmatic way in the text *On the Physician* of the Hippocratic collection. That is how it appears to have been understood by the medieval physicians, who, in imitation of the Hippocratic model, wrote treatises on the norms that should guide the activity of the physician so that the errors – diagnostic and therapeutic – do not damage the prestige of the profession. To this genre belong, for example, various Salernian texts, such as those entitled *Quomodo visitare debes infirmum* [16], *De adventu medici ad aegrotum* [17], and *De instructione medici secundum Archimataeum* [18]. Along these lines we can cite as well the text of Arnau de Vilanova, *De cautelis medicorum* [23], which is called *De circumspectione medici* in other manuscripts, as also another minor work called *Conditiones boni medici secundum Arnaldum de Villanova* [15]. All these texts have an ethical character, since they attempt to lay down guidelines concerning good professional practice and, more particularly, concerning good diagnosis. But all this is down within the limits of a concrete ethic, that of "social prestige", which places the good name and social esteem of the profession before any other principle. Since the social stratification is natural, and consequently good, the doctor considers that the preservation of his prestige is an ethical duty that must be safeguarded at all costs. This does not mean that the medieval doctor is not a religious man. He is. Yet, his religiosity serves to establish not the proximate norm of morality, but the ultimate norm. Thus, Arnau de Vilanova is a Christian and knows that God is love, charity. This charity, the love that God is, is the ultimate norm of mortality. But now then, this God has, precisely out of love, created nature and this natural order is the proximate norm of morality. To it belongs the order of the *civitas*, or the social order. Hence the physician's endeavor to preserve his prestige, i.e., the position that nature has assigned to him in the social order, is not only a natural and worthy desire, but also an ethical one; it is not only a desire, it is a duty.

B. The Religious Ethic of Medical Diagnosis in the Medical Diagnosis
  in the Middle Ages

The situation in the lower levels of medieval society is quite different. Strictly speaking, the technical or physiological medicine of the Hippocratic-Galenic tradition does not reach them. Hence the ethic that here obtains is not the physiological one of social prestige, but the religious one of care of the handicapped that we have seen in the first Christian communities. The poor are attended, not for reasons of prestige, but merely because of the love of God; that is to say, in this case a religious and not a physiological ethic is exercised. Thus, the care of the poor patient is usually based on the religious precept of love of one's neighbor, already present in the Hippocratic Oath. Form the second century A.D., this oath begins to gain more and more popularity among the physicians, and between the eighth and tenth centuries, it can be found in versions that are adapted to Jewish ([1] pp. 245–257; [19] pp. 440–454), Christian ([12]; [24]), and Muslim ([12] pp. 23–25; [13]; [14] pp. 173–203 [21] pp. 107–116) dogmata. The medieval physician attempts to combine the ethics of prestige with that of the good of the patient. The physician considers the care of the privileged strata of society a duty in justice imposed by nature. These people have to be attended to in justice and hence, always. The poor or people of the lower social strata are not attended to in justice, but because of charity and, therefore, in an intermittent, discontinuous, and sporadic way. On account of charity, the first Christian hospital was founded in Cesarea and, because of love for the poor, hospitals for the poor spread through Byzantium, Islam, and Latin Europe. Certainly, it appears that a certain technical assistance is due to the ill people in these places, but above all, they are given spiritual and sacramental help, the only thing which could always be given to those whom nature had denied any other type of aid.

While technical diagnosis in medieval medical practice among the upper levels of society so overshadows the ethical dimension that it almost annuls it, quite the contrary is the case in the care of the ill among the lower levels of society in the Middle Ages. Here the diagnosis becomes an ethical matter or more properly a religious one.

Complutensis University of Madrid
Madrid, Spain

NOTES

* The editors wish to express their thanks to Michael C. White for the translation of this essay.
[1] For further discussion of this point, see Bultmann [4], Wilkens [25], Schnackenburg [20], Fortna [8], Brown [3], and Dodd [5].

BIBLIOGRAPHY

1. Bar-Sela, A., and Hoff, H.E.: 1962, 'Isaac Israeli's fifty Admonitions to the Physician', *The Journal of the History of Medicine and Allied Sciences* 17, 245–257.
2. Boismard, M.E., and Lamouille,: A. 1977, *L'évangile de Jean* Edit. Du Cerf, Paris.
3. Brown, R.E.: 1966, *The Gospel According to John (I-XII)*, Doubleday & Company, Garden City, New York.
4. Bultman, R.: 1957, *Das Evangelium des Johannes*, Vandenhoeck Ruprecht, Göttingen.
5. Dodd, C.H.: 1963, *Historical Tradition in the Fourth Gospel*, Cambridge University Press, London.
6. Edelstein, L.: 1943, *The Hippocratic Oath*, The Johns Hopkins Press, Baltimore.
7. Edelstein, L.: 1965, 'The Professional Ethics of the Greek Physician', *Bulletin of the History of Medicine* 30, 392–418.
8. Fortna, R.T.: 1970, *The Gospel of Signs*, Cambridge University Press, London.
9. Gracia, D.: 1977, 'Etica y Medicina en el Cristianismo primitivo y en la Edad Media', in F, Gracía Miranda and J. de Arizcun Pineda (eds.) *Deontología, Derecho, Medicina*, Ediciones del Colegio Oficial de Médicos, Madrid, pp. 227–239.
10. Gracia, D.: 1980, 'Practica mesiánica y asistencia al enfermo', in A. Albarracín *et al* (eds). *Medicina e Historia*, Editorial de la Universidad Complutense, Madrid, pp. 355–387.
11. Gracia, D.: 1982, 'Diaita im Frühen Christentum', in H. Tellembach (ed.), *Psychiatrische Therapie heute*, Ferdinand Enke, Stuttgart, pp. 12–30.
12. Jones, W.H.S.: 1924, *The Doctor's Oath: An Essay in the History of Medicine*, Cambridge University Press, New York.
13. Levey, M.: 1977, Medical Deontology in Ninth Century Islam, in C.R. Burns (ed.), *Legacies in Ethics and Medicine*, Science history Publications, New York, pp. 129g–144.
14. MacKinney, L.C.: 1952, 'Medical Ethics and Etiquette in the Early Middle Ages: The Persistence of Hippocratic Ideals', *Bulletin of the History of Medicine* 26, 1–3.
15. Paniagua, J.A.: 1961, *El Maestro Arnau de Vilanova, médico*, Cátedra e Instituto de historia de la Medicine, Valencia, pp. 62–3.
16. Renzi, S. de (ed.): 1853, 'Quomodo visitare debes infirmum', in *Collectio Salernitana*, Tipografia del Filiatre-Sebezio, Napoli, II, p. 73.
17. Renzi, S. de (ed.): 1853, 'De adventu medici ad aegrotum', in *Collectio Salernitana*, Tipografia del Filiatre-Sebezio, Napoli, II, p. 74–80.
18. Renzi, S. de (ed.): 1859, 'De instructione medici secundum Archimathaeum', in *Collectio Salernitana*, Tipografia del Filiatre-Sebezio, V, p. 333–349.
19. Rosner, F.: 1967, 'The Physician's Prayer Attributed to Moses Maimonides', *Bulletin of the History of Medicine* 41, 440–454.

20. Schnackenburg, R.: 1971, *Das Johannesevangelium*, 2 Teil: John 5–12, Herder, Freiburg.
21. Schipperges, H.: 1965, 'La ética médica en el Islam medieval', *Asclepio* **17**, 107–116.
22. Schüller, B. 1976, 'Teoría Teleológica en la ética del amor', *Concilium* **120**, 540–545.
23. Sigerist, H.E.: 1946, 'Bedside Manners in the Middle Ages: The Treatise *De Cautelis Medicorum* Attributed to Arnald of Villanova', *Quarterly Bulletin of North Western University Medical School* **20**, 139–142.
24. Welborn, M.C.: 1938, 'The Long Tradition: A Study in Fourteenth Century Medical Deontology', in C.R. Burns (ed.), *Legacies in Ethics and Medicine*, Science History Publications, New York, pp. 204–217.
25. Wilkens, W.: 1969, *Zeichen and Werke. Ein Beitrag zur Theologie des vierten Evangeliums in Erzählungs und Redestoff*, Zwingli Verlag, Zürich.

AGUSTÍN ALBARRACÍN

THE ETHICS OF DIAGNOSIS IN THE MODERN AND CONTEMPORARY WORLDS*

I. INTRODUCTION

Transformations that have taken place in the modern epoch have brought about a drastic change in the religiosity and fundamental beliefs of Europeans, including, of course, physicians. This change provoked in its turn a division between religious and civil duties. From the sixteenth to the eighteenth centuries, a progressive secularization occurred in which the stage of Christian modernity may be distinguished from that of secularized modernity. From the eighteenth century onwards, with the secularization of the modern world, a diversity of attitudes toward the duties imposed by medicine arose alongside the Christian ethics. New critical and revisionist attitudes towards ethics emerged, and with it the demand for a rational basis for morality. This essay explores the influences of these attitudes on medical ethics and points out how the notion of ethics changed as one moved from the sixteenth to the twentieth century ([2] pp. 241–247; [4], pp. 15–31).

II. THE ETHICS OF DIAGNOSIS IN CHRISTIAN MODERNITY (16TH-18TH CENTURIES)

Christian religious morality, both in its Catholic and Protestant expressions, was still "Hellenized", just as in the Middle Ages, and so the moral norm remains the "natural order", as established by God the Creator of nature. Consequently, the ethical schemes proposed by Professor Gracia ([14] pp. 227–239; [15], pp. 22–25) for the Middle Ages are still valid for Christian modernity. However, during the sixteenth to the eighteenth centuries, the state established both civil and religious duties and imposed them on physicians. The physician was thus confronted with three different types of obligations: those arising from his own conscience, the truly ethical ones; the religious ones prescribed by the secular authorities; and the purely civil ones.

Between religion and medicine, both a theoretical and a practical relationship existed. The 'right' knowledge of the world was held to derive

29

J. L. Peset and D. Gracia (eds.), The Ethics of Diagnosis, 29–39.
© *1992 by Kluwer Academic Publishers. Printed in the Netherlands.*

from two sources: the observation of natural reality, the expression of God's creation; and the exegesis of the divine word through the Scriptures. The expression par excellence of this spiritual approach is found in the work of eighteenth-century physician Friedrich Hoffmann, (1660–1742), *Medicus Politicus* [17], which carries the subtitle: "Rule of Prudence according to which a young doctor should approach his studies and his way of life". In the author's view, one of the first of these rules is that "A doctor must be a Christian" ([17], Regulation VII). Faithful to this maxim, Hoffman in his booklet *Dissertation Theologico-Medica de officio boni Theologi ex idea boni Medici* ([18], pp. 384–422) states that the physician's total knowledge must be completely and substantially informed by Christian truths. As both theologian and doctor, Hoffmann resorted to two different and concurrent ways of knowing the world – and hence of knowing disease – namely, direct *observation* of natural reality, and *exegesis* of the sacred texts relevant, more or less, to visible nature.

This attitude gave rise, however, to a series of conflicts in the Christian physician when engaged in the diagnosis of patients. Let us turn to two paradigmatic examples, one prior and the other posterior of Hoffmann's thought.

The seventeenth Century physician and Puritan, Thomas Sydenham (1624–1689) was along with his colleagues and friends in the Royal Society, committed to the Baconian method. This method proclaims that to know the reality of the disease "one need not imagine nor solve, but discover what Nature does and produces" ([31], p. 144). Indeed, this discovery – via the diagnostic act – implies the necessity of experimental observation and an attempt to explore instrumentally the essence of reality. Sydenham's theological commitments contrast with the practices of the Baconian method. The physician's *duty* consists of limiting his observations to "the outer husk of things" ([31], p. 88), since God regulated his faculties to the sole perception of the surface of bodies and not to the hidden and intricate processes of nature's "abyss of cause" ([31]; [32], p. 210). As a consequence, the study of pathological anatomy in general, and the use of the microscope in particular, must be proscribed for religious reasons. Sydenham's "ethical" attitude led him to the discovery of the modern concept of "species of disease". In describing diseases, he attended to the "superficial" or external consideration of symptoms and did not integrate the "hidden and intricate" essence of the morbid process. But his modern concept of "species of disease" ([1]

pp. 297–307) did not lessen the reality of the conflict between medical knowledge and his religious beliefs, and the practical duties derived from both.

In contrast, one might consider Réné-Théophile-Hyacinthe Laennec (1781–1826), a representative of the Christian physician of the nineteenth century. Faithful to an anatomo-clinical conception of illness, Laennec views diagnosis as the art of knowing *in vivo*, through physical signs, the organic lesions in the patient's body. To be sure, the discovery and determination of physical signs demanded the practice of exploratory maneuvers, e.g., thoracic percussion and immediate auscultation, which had then been recently introduced into clinical medicine by Jean-Nicolas Corvisart (1755–1821) and Gaspard-Laurent Bayle (1774–1816). Confronted with a young, slightly obese, ill woman, however, the clinician Laennec become the Christian doctor as he asks: Is it morally licit to bring the ear into contact with the ill woman's thorax ([20], p. 10)? Again we witness the conflict between medical knowledge and religious beliefs even though something as important for the future of diagnosis as the discovery of the stethoscope arose from this situation.

Granted, these are extreme cases. With the secularization of the modern world, the Christian physician by and large acted toward the patient as if medicine, insofar as it is a science, had no relation with religion. Insofar as medicine is a practice, however, it was thought to hold a relation with religion. Some bridge had to be laid between a natural science that was alien to divinity, and a purely personal religious piety, (a Christian medical deontology understood as a pure system of practical rules). So, a division between civil and religious duties was recognized: the latter were relegated to the Christian moral conscience and to the laws that govern it, while the former were prescribed by the secular authorities on the basis of a new axiological system.

But when a professional ethics clashed with a theological ethics, how was the physician to resolve the conflicts encountered in practice? Would it not be, as in the Middle Ages or in the primitive Christian communities, the personal conscience of the physician that should decide?

III. THE ETHICS OF DIAGNOSIS IN
SECULARIZED MODERNITY

The secularized man of the eighteenth century sought a basis for morality in rationality, not religious ideology. Influenced by Grotius and the works of the British moralists (from Thomas Hobbes (1588–1679) and John Locke (1632–1714), to David Hume (1711–1776) and John Stuart Mill (1806–1873) who in turn influenced Benedict Spinoza (1632–1677), Immanuel Kant (1724–1804), Georg Wilhelm Friedrich Hegel (1770–1831), Friedrich Nietzsche (1844–1900), and others), one attempted to fashion principles and theories that could justify and give value to secular ethics, and thus provide the basis of a new concept of duty. As a result of this historical development, contemporary health care professionals practicing in secular societies came to establish their own professional ethics and these provide, in part, the basis of the diagnostic approach.

Four values are basic to a secularized ethics of diagnosis: altruism, social welfare, a religion of the facts, and prestige ([21] pp. 293–294 and 383–384).

A. *Altruism*

Altruism – a term introduced by Auguste Comte (1798–1857) ([9], I, p. 693; IV, p. 289) – can be seen as a principle involving the secularization of love toward the neighbor under the form of a philanthropy that forms the basis for technical and care activity. As such, it is a movement of projection toward the I of the other, which impedes the natural impulses of self-love and which should, in a positivist society, necessarily overcome such impulses, and constitute itself as the moral basis of the new society.

Confronted with a patient, the altruistic doctor diagnoses his complaints. Granted, there is a primary movement of pleasure in intellectual clarification in response to the challenge that the patient's body offers the physician. Fundamentally, however, the physician engages in the act of diagnosis because the solution of the diagnostic problem and its subsequent treatment mean for him a behavior professionally regulated by an oscillating commitment between the legal rules of Hegelian "objective spirit" and the inner commands of Kant's "categorical imperative". Even though the physician may proclaim himself to be neither Hegelian nor Kantian but positivist and agnostic, diagnostic exploration for the most part proceeds according to the above described ethical conduct. In this, contemporary European and American physicians have followed altruistic ethics, i.e, that of the patient's welfare, as a primary approach that demands a professional morality authentically exemplary, and in many cases more exquisite and demanding than that of the physician who confessed to be sincerely Christian.

B. *Social Welfare*

Some might argue that the good of society comes before that of the patient, although the latter naturally conditions the former. If one considers all citizens as co-operative members of society, as links in a complex machinery, the society's welfare is an obvious presupposition for the smooth economic and political organization of the community. In the transition from feudal to bourgeois life, the division of labor imposed by the new socio-economic structure brought about new diseases conditioned by this new way of life. For further elaboration of this point one might consider, for example, Paracelsus (1493–1541), *Von der Bergsucht* (1533–34) [24], Bernardino Ramazzini (1633–1714), *De morbus artificum* (1700) [27] and Johann Peter Franck, (1745–1821) *System einer vollstandigen medizinischen Polizey* (1779–1789) [13]

In addition, the nineteenth century physician Max von Pettenkofer (1818–1901) [26] was to convert hygienics into one of the branches of applied natural science and statistics, and, along with Sir Edwin Chadwick (1800–1890), indicate the influence health had upon the so-called labor force. Recall Pettenkofer's remark:

Let us fix the amount of expense and losses for each day of illness at one florin, a figure which we consider to be minimum and far below the average. If the mortality rate in Munich only descends from 33 to 30 per thousand, what is the minimum value of money corresponding to that reduction?...The annual reduction of 17,340 clinical cases represents 346,800 days of illness, which is equivalent to the same number of florins if we calculate the daily loss at one florin ([26], pp. 318–320).

Von Pettenkofer directed his calculations toward support for urban sanitation measures as a prophylaxis of disease. But *mutatis mutandis*, could a physician whose vocation is the community good fail to take into account the saving of a florin per day when diagnosing his patients? Would he not be acting under a socio-economic and ethical imperative in establishing the diagnosis?

C. *The Religion of Facts*

The ethics founded on Comte's philosophical system, positivism, assigned priority to the role of facts in accounts regarding the observed world. The introduction of the "positive method" in nineteenth-century pathology contributed in part to the significant development of the concept of 'disease'. The implications of the acceptance of the "positive method" by such physicians as Marie-Francois-Xavier Bichat (1771–1807) and Claude Bernard (1813–1878), particularly in the realm of medical diagnosis, were not entirely beneficial. To use Ramon y Cajal's expression, these physicians, as "uncompromising fanatics of the religion of facts" ([28], p. 287) made a "fetish of the concrete" ([12], p. 428). In doing so, the values of altruism and the social welfare were relegated to positions subordinate to pursuit of concrete facts, whether anatomic lesions, dysfunctions, or specific bacterial infections ([3], p. 14).

This obviously posed a problem. Up to what point does the objective scientific, or natural consideration of the patient compel the doctor to act ethically when it comes to establishing the diagnosis of the disease in the concrete person who suffers it? We have already seen Laennec's response to such a case. But the historical reality compels us to admit that during the nineteenth and the first half of the twentieth centuries, medical ethics did not attend to the types of problems that stem from a traditional paternalist conception of the physician-patient relationship, in which the figure of the physician was exhalted and the patient assigned a passive role.

D. *Professional Prestige*

Consider now the least noble form of diagnostic behavior with respect to the patient. Professional prestige is expressed in the physician's egoism, which is equivalent to "self-interest". There has been a professional rivalry among the multiple and diverse medical degrees that resulted from the profusion of faculties vis-a-vis the patient. One must remember the simple fact that, in 1862, no fewer than thirty-five different kinds of medical professionals converged upon Spain – giving rise throughout the modern period to attempts, fundamentally egoistic although quite justified, to cultivate an ethic of prestige that is demonstrated by the investigative and technical capacity of the healer. We could denominate this approach as the ethic of success. Its raison d'etre is found in the competitive character of bourgeois society, which supports not only cultivation of prestige but the profit motive.

This approach has posed a series of moral conflicts from the very outset of modernity. Initially, exploration of the patient was limited to some meager maneuvers such as sensorial apprehension, inspection of urine, and to subtle qualification of the pulse. The more informed physicians understood that these maneuvers were of little importance in determining the real essence of the disease. The French Samuel Sorbicre (1615–1670), for example, exposed his doubts in this respect in his *Avis a un ieune medecin sur la maniere dont il se doit comporter en la pratique de la Medecine,* Lyon (1672):

Urine is but a serous liquid inpregnated with salts filtered from the kidney which goes to the bladder and which can indicate the condition of those parts... And with it, nevertheless, one is supposed to be able to conjecture about the state of the rest of the body and to know what is going to take place in the body. This method of observing the urine has given rise to thousands of jokes made upon those doctors who have tried to express a judgement by means of such an unreliable method ([23], p. 262).

The criticism could not be more to the point. However, in a paragraph just previous to the above, Sorbiere clearly lays out the ethics of prestige:

In order to adjust oneself to the ignorant public's credulity, curious to know the conditions and results of a disease, it has been necessary to create a fiction of obtaining knowledge from the inspection of the urine far beyond what could be discovered from such an inspection... ([23], p. 262).

The intent here is obvious: the physician, quite apart from his own professional and personal conscience, knows the uselessness of the diagnostic procedure and needs to show a faction of knowledge to satisfy the public's demand.

In keeping with the tradition of scientific medicine, Sorbiere also devoted some chapters of his booklet to the conduct of the respectable physician, to what he must do and say vis-a-vis the patient. This is a tradition that was very quickly developed further (recall here Hoffmann's booklet) and which led to the publication in the nineteenth century of Codes of Medical Ethics.

IV. THE MODERN WORLD: FROM ETHICS TO DEONTOLOGY

In light of the previous discussion, one can conclude that regardless of the physician's orientation, whether Christian, philanthropic, socially active, or motivated toward personal prestige, it is a physician's conscience that ultimately determines the rectitude or moral integrity of medical practice. Be this as it may, the modern world could not leave professional norms to the utter discretion of the personal conscience of the practicing physician. As a consequence, it was necessary to take the step from Kantian *Moralität* (the doing of one's duty by an act of will), to the *Sittlichkeit* (obedience to the moral law as established by the norms, laws and customs of society, which in turn represents, to Hegelian language, the objective spirit or one of its forms). In other words, it was necessary to make the transition from ethics to deontology.

In my view, the so-called codes of ethics first arose with a clearly determined goal, and it was only afterwards that their contents were enlarged. That goal was the regulation of the political and scientific relations among physicians, which were strained by the already mentioned proliferation of degrees and faculties. One might argue that the real aim of those codes, was the enhancement of the prestige of the profession rather than the formulation guidelines regarding "good" behavior in medical practice.

   Although many hold the first code of ethics to be the work of Thomas Percival of Manchester (1740–1804) ([5] pp. 52–58; [6]; [11]; [19], pp. 165–166), there is evidence to the contrary. In 1772, the Edinburgh physician, John Gregory (1724–1773) published six lectures [16] that were part of a course given two years earlier on the physician's qualities and duties. This appeared well before the 1803 publication of Percival's *Medical Ethics* ([8] p. 300). In any event, an analysis of both texts enables one to understand their common root, namely, the desire to stress the ethical values of medical education, of the interprofessional and physician-patient relations, and of the relations that should link the physician with the community. This is not to say that one cannot find a distinct differences in their "ethics of prestige". Following Berlant [5], who has carefully studied this theme, one can see that while both authors were enlightened, the liberal Gregory based medical prestige on the physician's scientific formation, whereas the conservative Percival stressed the monopoly of an *esprit de corps*. Gregory's ideal presupposed the acceptance of competition. Not only physicians, without distinction of faculties, have a duty of formal education as the ethical basis for a good practice, but so also do enlightened gentlemen, although in the manner of a layman, who should devote his attention to the basic problems of illness and health. Evidently, Gregory was aware of the spirit informing Adam Smith's *Wealth of Nations* [30], published in Great Britain at that time.

   Against this anti-monopolistic attitude, Percival's *Medical Ethics* – conceived as a mediatory instrument in an epoch in which physicians, surgeons, and apothecaries struggled with one another – conservatively defended professional monopoly. Hence Percival's greater dedication was to the study, not of educational themes, but of interprofessional relationships in hospitals and private or general practice, and of the conduct of physicians vis-a-vis apothecaries. This may be seen as an ethical code written by physicians and addressed to physicians, where the question of the ethically good diagnosis (an objective to which Gregory's competitive approach led in as much as it was concerned with the medical prestige and scientific formation of physicians) is of lesser importance than the question of the regulation of interpersonal relationships. By rigorously maintaining the class spirit, the result would be the restriction of competition by eliminating those without professional licenses. At the same time, professional errors could be silenced and public criticism prevented. Percival adapted the guild regulations of the London Royal College of Physicians to the framework of hospital and private practice. The prestige of the medical class was to be the basis of his *Medical Ethics*.

The works by Percival and Gregory, together with Michael Ryan's (1800–1841) *Manual of Medical Jurisprudence* (1831) ([7], p. 301) constitute the foundation for the codification of medical ethics in the English speaking countries. All Anglo-American codes, beginning with the *Sixteen Introductory Lectures* by Benjamin Rush 1745–1813 ([29], pp. 141–165) and the ethical code of the American Medical Association (1847), combine Gregory's liberalism and Percival's conservatism, thus offering norms of morality that regulate the relationships among physicians themselves and between physicians and patients. Such codes reflect the problems of intrusionism, quackery, and of the still undefined medical specialization. They are, as I have already said, codes elaborated by physicians for the use of physicians in very restricted realms, and only very indirectly bear on the ethics of diagnosis. In Europe, the same is true of the different professional statutes drawn up by physicians for the regulation of their practice.

Centro de Estudios Históricos, C.S.I.C.
Madrid, Spain

NOTE

* The editors of this volume are indebted to Michael C. White for the translation of this essay.

BIBLIOGRAPHY

1. Albarra cín, A, 1973, 'Sydenham', in P. Laín Entralgo (ed.) *Historia Universal de la Medicina*, Vol, *IV*, Salvat Editores, Barcelona, pp. 297–307.
2. Albarracín, A.: 1977, 'Deontologia médica en el mundo moderno y contemporaneo', in Colegio Oficial de Medicos (ed.) *Deontologia, Derecho, Medicina*, Madrid, pp. 241–247.
3. Albarracín, A.: 1980, 'Cajal y el problema del conocimiento cientifico', in *Asclepio, Archivo Ibero Americano de Historia de la Medicina y Antropologia Medica* 32, pp. 7–17.
4. Albarracín, A.: 1982, 'Historia de la etica en Medicina', in M. L. Marti (ed.) *Etica en Medicina*, Fundacion Alberto J. Roemmers, Buenos Aires, pp. 15–31.
5. Berlant, J.L.: 1977, 'From Medical Ethics and Monopolization', in S.J. Reiser *et al.* (eds.), *Ethics in Medicine. Historical Perspectives and Contemporary Concerns*, M. I. T. Press, Cambridge, Massachusetts, pp. 52–65.
6. Brand, U.: 1977, *Ärztliche Ethik im 19 Jahrhundert*, Hans Ferdinand Schulz Verlag, Freiburg i.B.
7. Burns, C.R.: 1977, 'Reciprocity in the Development of Anglo-American Medical Ethics, 1765–1865', in C.R. Burns (ed.), *Legacies in Ethics and Medicine*, Science History Publications, New York, pp. 300–306.
8. Burns, C.R.: 1977, 'Thomas Percival: Medical Ethics or Medical Jurisprudence?', in C. R. Bums (ed.) *Legacies in Ethics and Medicine*, Science History Publications, New York, pp. 284–299.
9. Comte, A.: 1851–1854, *Systeme de politique positive, instituant la religion de l'Humanité*, 4 vols., E. Thunot et C. Paris.
10. Dewhurst, K.: 1966, *Dr Thomas Sydenham (1624–1689). His Life and Original Writings*, University of California Press, Berkeley.
11. Duncan, A.S., Dunstan, G.R. and Welbourn, R.B. (eds.): 1977, *Dictionary of Medical Ethics*, Darton, Lingman and Todd, London.
12. Duran, G. and Alonso Buron, F.: 1960, *Cajal. I. Vida y Obra*, Institución "Fernando el Católico", Zaragoza.
13. Frank, J.P.: 1779–1788, *System einer vollstandigen medicinischen Polizey*, 4 vols., C.F. Schwan, Mannheim.
14. Gracia, D.: 1977, 'Etica y medicina en el cristianismo primitivo y en la Edad Media', in Colegio Oficial de Medicos (ed.), *Deontologia, Derecho, Medicina*, Madrid, pp. 227–239.
15. Gracia, D.: 1992, 'The Ethics of Diagnosis in the Early Christianity and the Middle Ages', in this volume, pp. 19–27.
16. Gregory, J.: 1772, *Lectures on the Duties and Qualifications of a Physician*, Straham and T. Coddel, London.
17. Hoffman, F.: 1738, *Medicus politicus, sive regulae prudentiae secundum quae medicus juvenis studia sua et vitae rationum dirigere debet, si famam sibi felicemque praxin et cito acquirere et conservare cupit*, Apud Philippum Bank, Lyons.
18. Hoffman, F.: 1754, *Operum omnium physico-medicorum supplementum*, Fratres de Tournes, Geneva.
19. Konold, D.E: 1978, 'Codes of Medical Ethics: History', in W. T. Reich (ed.) *Encyclopedia of Bioethics*, Vol. **I**, The Free Press, New York, pp. 162–171.

20. Laennec, R. T. H.: 1837, *Traite de l'auscultation mediate et des malladies des poumons et du coeur*, 4th ed. Vol. I, J. S. Chaude, Paris.
21. Laín Entralgo, P.: 1964, *La relacion medico enfermo. Historia y teoria*. Revista de Occidente, Madrid.
22. Leake, C.D.: 1975 *Percival's Medical Ethics*, Robert E. Krieger Publishing Company, Huntington.
23. Lester Pleadwell, F.: 1977, 'Samuel Sorbiere and his Advice to a Young Physician' in C. R. Burns (ed.), *Legacies in Ethics and Medicine*, Science History Publications, New York, pp. 237–269.
24. Paracelsus, T. V.: 1533–1534, *Von der Bergsucht und anderen Bergkranheiten*, Durch Gebaldum Mayer, Diligen, in K. Sudhoff (ed.): 1922–1933, *Teophrast von Hohenheim, gen. Paracelsus, sämt. Werke*, 14 vols., Oldenbourg, München, Vol. **IX.**, 461–544.
25. Percival, T.: 1803, *Medical Ethics or a Code of Institutes and Precepts, Adapted to the Professional Conduct of Physicians and Surgeons*, S. Russell, Manchester.
26. Pettenkofer, M.v.: 1873, *Über den Wert der Gesundheit fur eine Stadt*, Vieweg, Braunschweig. Spanish trans. J. M. Lopez Piñero, *Medicina, Historia, Sociedad*, Ariel, Barcelona 1969.
27. Ramazzini, B.: 1700, *De morbis artificum diatriba*, A. Capponi Modena.
28. Ramon y Cajal, S.: 1923, *Recuerdos de mi vida*, 3rd ed., Imp. J. Pueyo Madrid.
29. Rush, B.: 1811, *Sixteen Introductory Lectures*, Bradford Innskeep, Philadelphia.
30. Smith, A.: 1776, *Inquiry into the Nature and into the Causes of the Wealth of Nations*, W. Coller, Dublin.
31. Sydenham, T.: 1735, 'Tractatus de Podagra et Hydrope' and 'Observationes Medicae circa morborum acutorum historiam et curationem', in *Opera Medica* Balleoniana, Venice.
32. Wolfe, D. E.: 1961, 'Sydenham and Locke and the Limits of Anatomy', *Bulletin of the History of Medicine* **35** (3), 193–220.

JOSÉ LUIS PESET

MEDICAL DIAGNOSIS AND INSTITUTIONAL SETTINGS*

The ethics of diagnosis cannot be studied independently of the moral values of the group or class to which the physician and ill person belong. And these values are determined by the epoch and the place where they are lived. Only in this way can we comprehend the historical evolution of medicine. Pedro Laín Entralgo [3] and Michel Foucault [2], among others, have explored this basis of continuous change in the physician-patient relationship.

In pretechnical or supernatural medicine, physican and patient were separated by barriers of superstition and fear. Barriers were erected through the social stratification of castes [1]. A small caste of priests with enormous social and religious power administered medicine. Ill persons were not patients but sinners who approached the temple in search of pardon for their sins. It is the Greek physician, a free and scientific physician, who broke these barriers. As a consequence, medicine entered a framework of rational and free interchange with the patient. The Hippocratic physician, like the hero Prometheus or the fabulous centaur Quirinus, approached the ill person, directly investigated the problem that ailed him, applied his art, and charged his fees. Contemporary medicine was thus born in the century of Pericles [4].

Modern medicine, after the lapse of the long medieval period, follows the paths opened up by the author of *De morbo sacro*: observation of the disease or illness converted into a natural reality, close contact between the physician and the patient in hospital or private institutions, a direct commercial relationship between the artisan and the consumer of that artisanship or skill. But with the onset of capitalism and liberalism, each one of those characteristic notes of modern medicine carried within itself an ancient germ of discord, an old motive of separation between the doctor and the patient. The scientific study of illness led to the deprivation of the patient's primary attribute, namely his or her illness. Species of disease were classified, described, and reified; and the ill person was once more a case which realized an abstract complex. The development of modern techniques of diagnosis further aggravated this process of

41

J. L. Peset and D. Gracia (eds.), The Ethics of Diagnosis, 41–45.
© 1992 *by Kluwer Academic Publishers. Printed in the Netherlands.*

fragmentation, separation, and objectification of the ill person, cutting him or her off from social and individual reality.

The new institutions, especially the hospital and the medical specialist contributed to this process. The patient was isolated from the physician, from society, and from his or her ailment. The patient became a number, a clinical case history, a punched card; the physician was transformed into a complex Kafkaesque mechanism phenomenalized for the patient by the large number of people without identity, by colorless rooms, and by meaningless apparatus. The highly complex pluripersonal and pluritechnified process of diagnosis gave rise to a separation between physician and patient quite similar to that which occurred in primitive cultures. Frequently the patient's fear of these was no less; the physician's lack of interest in the ill person's disease process was also marked by distance.

The commercial relationship between physician and patient is likewise marked by serious disintegration. The incisive mockery of the sharp-witted Artistophanes towards those Greek physicians who did not exercise their medical activities without the quest for rapid personal economic gain, as well as that of Quevedo and of Molière, all suggest the problem. The quest for lucre, rooted in the new capitalist system, carried grave consequences, contributing to the inevitable concentration of capital and the lamentable proletarization of the working class. And so, in fact, unchecked increase in medical costs was not only a consequence of a clear technical and therapeutic improvement, but also a consequence of the enormous profitability of economic investment in relation to medicine and health. Besides, medicine plays a primordial role when it comes to the relationship between capital and the labor force. The different social security systems were established to maintain the strength and peace of the reserve armade of workers who supply labor for capitalism.

Here is a theme which deserves greater attention: the origin and motives for establishing security against illness including the systems of collective medicine. The cost of medicine, the painful effect of the crises of capitalism, and workers' claims continually oblige the capitalist to make fresh concessions, in this instance a medical system which can assure more, just and better medical care for the working class. But this system falsifies the true conditions of a social medicine, because it is indirectly aimed towards bettering and consolidating the mode of production. Moreover, each national system of social security presents its own peculiar characteristics. In the Spanish case, one could point out a deficient administration, a lack of say in management by truly interested parties, inadequacy in primary medical care, serious inadequacies in special, technical and economic coverage, among other things. Even the most highly developed system of social security, Britain's National Health Service, suffers these defects especially since the grave economic crisis of 1973. The British Health Service is today seriously troubled by economic, technical, and personnel problems, and recent attempts at improving the administration and management unfortunately have not been able to solve them [5].

Within the framework of these institutional aspects I should like to add, or at least refer to, another institution, which in my mind markedly determines the type of physician-patient relationship. I mean the medical faculties, or those institutions which throughout the history of medicine have been responsible for training physicians. The esoteric role of the temples in the teaching of medicine to the priests was quite different from that played by the different Greek schools of medicine. The latter, much more open, much more scientific, enabled modern medicine to come into being. The same may be said of the Toledo school of translators or the school of Salerno in the role of transmitters of classical culture to medieval Europe. The university with its faculties of medicine meant once more a step towards the archaic and the sacred in medical knowledge which the medical teaching of new schools and hospitals managed to break. The role played by the hospitals of Leiden, of Vienna, or of Edinburgh in the creation of modern medicine was extraordinary. The approach to the ill person's bedside which modern Europe came to know thanks to the efforts of Boerhaave and his disciples, led to the birth of a naturalistic, observational, and classificatory medicine. But in the long run the teaching of medicine suffered the same deterioration as medical care. Both in the French model and in the German one – that is, in the clinical model and in the scientific one, the patient was soon turned into an object, all too soon dissected, objectified, and depersonalized to serve as a mere example of a clinical case history. It is not surprising that the great nineteenth-century clinicians such as Skoda or Addison ([3], pp. 204–210) frequently overlooked consigning or advising treatment. That was not what was being pursued; rather, the great physicians of those hospitals were interested in the patients for the construction of a great medico-sanitary edifice. The ill person himself, in his relationship with his illness and with his professional and social status, was not an object of interest. Today those same defects continue to arise. The teaching of medicine, at least in many countries, including Spain, suffers from a great excess of scientism and specialization, with little concern about the truly personal, economic and social roots of the process of illness.

I should not like to omit mentioning the importance of epidemic diseases. From the beginning of time this type of collective disease has had a special characterization. In the pretechnical world it was characterized as the punishment of the gods. Even in the classical world it seems to elude an essentialist interpretation, inasmuch as the Greeks and Romans dwell more on the characterization of its effects rather than its essence. Since then, however, epidemics were the first diseases to be the object of a therapeutic approach of a collective, non-individual type. Important advances in medicine, especially in questions of private and public hygiene, were due to them. The forms of collective medicine owe a great deal of their innovations to encounters with epidemic diseases. Consequently, it seems to me appropriate to dwell on epidemiological problems when dealing with an analysis of institutional, historic, and social conditioning factors of disease or illness and its treatment. Nowadays, one may say, following Johann Peter Frank, that a country can measure its medicine in relation to the quality of its epidemiological and sanitary organization [6].

Finally, I should like to underscore that only by furnishing the analysis of ethical values with a historical, social, institutional and economic dimension and approach will it be possible to evaluate the role which medical ethics has had in the regulation of physician-patient, physician-society relations. The physician and his patient are never on a desert island, as the personal physician of Bismark maintained, rather, they are found in a determinate place and time ([3], p. 20). It is only from the correct analysis of those circumstances which determine medical diagnosis and treatment that any sensible interpretation of medical ethics could arise.

Centro de Estudios Históricos, C.S.I.C.
Madrid, Spain

NOTE

* The editors wish to thank Michael C. White for the translation of this essay.

BIBLIOGRAPHY

1. Albarracín, A.: 1970, *Homero y la medicina*, Prensa española, Madrid.
2. Foucault, M.: 1963, *Naissance de la clinique*, P.U.F., Paris.
3. Laín Entralgo, P.: 1964, *La relación médico-enfermo*, Revista de Occidente, Madrid.
4. Laín Entralgo. P.: 1970, *La medicina hipocrática*, Madrid.
5. Peset, J.L.: 1978, 'Capitalismo y medicina: ensayo sobre el nacimiento de la seguridad social', *Estudios de Historia Social* **7**, 185–216.
6. Peset, J.L. and M.: 1972, *Muerte en España*, Seminarios y ediciones, Madrid.

DARREL AMUNDSEN

SOME CONCEPTUAL AND METHODOLOGICAL OBSERVATIONS ON THE HISTORY OF ETHICS OF DIAGNOSIS*

The medical morality of any society is, for the most part, roughly congruous with the broader moral perceptions of that society. Medical ethics, as a subset of any society's ethics, may be as varied as the society's ethics is pluralistic and as muddled as the society's ethics is confused. Whenever and wherever that which can be called medicine (even if it is a magico-religious medicine) is practiced by individuals who function as healers in society (even if their healing role is only one of several of their functions), there is an ethical framework in which such healing activities occur. This is true even in societies in which there was no medical literature, much less a distinct genre of medical ethics. But most of the diverse issues that today constitute bioethics first arose in the literature only when social, religious, philosophical, or economic change or scientific and tehnological advances created new moral conundrums or highlighted either certain ethical presuppositions or operative ethical strictures that require examination and clarification. Obviously the development and implementation of new medical technology have created and continue to create ethical dilemmas. Furthermore, new bioethical categories are sometimes created when aspects of medical epistemology or procedures are first examined with a view to their ethical implications or are examined afresh.

In the 1960s, an unprecedented bioethics movement began in the West that has been gaining momentum ever since. This movement was generated and stimulated by a variety of factors. Developments in medical technology dramatically increased public interest in and concerns about some aspects of medical ethics. Additional catalysts appeared. Burgeoning demands for consumers' rights coupled with the civil rights movement stimulated a patients' rights movement. The feminist movement and the "new morality"/sexual revolution brought the issue of abortion into public fora for discussion and debate. Liberalized abortion laws not only increased the intensity of the debate but raised public awareness of and sensitivity to claims of the "right to live," and the "right to die",

47

J. L. Peset and D. Gracia (eds.), The Ethics of Diagnosis, 47–61.
© 1992 *by Kluwer Academic Publishers. Printed in the Netherlands.*

which are made continually more perplexing by increasingly efficient medical technology. Bioethical issues continue to receive attention from the media, to provide the basis for litigation, to be the focus of much political debate, to fan the fires of religious fervor, indeed to be discussed by nearly everyone from the man-in-the-street to that byproduct of the bioethics movement itself, the bioethicist.

The quantity of literature generated by the bioethics movement is enormous and the variety of topics subsumed under the rubric of bioethics ranges from the most minutely specialized considerations to efforts to develop broadly based philosophical systems of bioethics. But at any time in human history, whether before or during the current bioethics movement, has ethics of diagnosis ever been isolated for special systematic scrutiny as a circumscribed ethical category?

The single most significant scholarly product of the bioethics movement is *The Encyclopedia of Bioethics*, published in 1978. Its purpose was, in the words of its Editor-in-Cheif, Warren T. Reich, "to synthesize, analyze, and compare the positions taken on the problems of bioethics, in the past as well as in the present, to indicate which issues require further examination, and to point to anticipated developments in the ethics of the life sciences and health care" ([7], p. xv). Reich recognized that it was "unusual, perhaps unprecedented, for a special encyclopedia to be produced almost simultaneously with the emergence of its field..." ([7], p. xvi). Two reasons

persuaded us that this work was not premature. First, many of the issues and methodological roots of bioethics are not new; they were waiting to be gleaned from centuries of literature in the fields of philosophy, medical ethics, history of medicine, and other fields. Second, the plethora of comtemporary literature badly needed systematization... ([7], p. xvi).

In organizaing this process of synthesizing, analyzing, and comparing both past and present positions, of indicating which issues would require further examination, and of pointing to anticipated developments in bioethics, the editorial board displayed both perspicacity and ingenuity in making the encyclopedia broad and inclusive. It certainly must be of some significance that "diagnosis" is not an entry in the *Encyclopedia of Bioethics*. It is also revealing that H. Tristram Engelhardt's recently published *The Foundations of Bioethics*, a searchingly analytical and exceedingly detailed study by a scholar who is intimately versed in the history of medical ethics, not only does not include any section or sub-section dealing with ethics of diagnosis, but does not even include diagnosis in its thorough index. It appears that ethics of diagnosis has not yet been isolated for special, systematic scrutiny as a circumscribed ethical category.

How, then, may the historian construct a meaningful history of ethics of diagnosis? Historians have occasionally sought to contribute some insights to the current dialogue on bioethics by writing limited or broad histories of medical ethics. Such efforts are quite varied in approach. 1) Some historians of medical ethics refuse to make the available sources address questions of current bioethics unless these were matters about which the primary sources were also concerned. 2) Others address questions to their sources that could have been raised before but apparently were not. 3) Yet others do not hesitate to address issues that could not even have been posited by their sources but reflect current suppositions for the philosopher *qua* historian or the historian *qua* philosopher. For example, if bypass surgery and kidney dialysis had been available in fifth- and fourth-century B.C Athens, how might Plato's strictures on euthanasia and the role or physicians in his *Republic* have been affected? Or if current life support systems had been available, how might the "Hippocratic" physician have applied his principle, "To help or at least to do no harm"?

In the second approach one encounters discussions of questions that could have been raised but either were not raised or evidence of their consideration has not survived. Here the historian must be very careful, for it is exceedingly easy to read issues of today into a past environment where they would never have arisen. For example, in a society that grants a monopoly to a group licensed to practice medicine, numerous ethical issues may arise concerning professional deontology and patients's rights, which would not be considered in a society in which the practice of medicine was laissez-faire. If one addresses to the latter questions that are quite appropriate for the former, one's answers may reveal little aside from the investigator's own interests and prejudices.

The first approach is the safest and most historically tidy. Numerous ethical issues have been raised and discussed in the genre of medical ethics for over two and one half millennia. Questions of fees or responsibilities of physicians vis-à-vis the terminally-ill are examples of ethical problems that were regarded as significant during many eras in the past. The historian can talk intelligently about these matters, especially if he discusses them within the broader context of social and intellectual

history. The historian may also work responsibly within this category by
addressing ethical issues not raised in the literature of medical ethics *only*
if relevant evidence appears in non-medical sources. A nearly trivial
example: The question of medical treatment of close relatives or spouses
appears not to have been raised in the medical literature of classical
antiquity. Sufficient evidence is available to demonstrate that the treatment of close relatives or spouses was not unusual. Hence, in the absence
of any questions having been raised about either the wisdom or the ethics
of such practice, one may safely conclude that it had not yet become an
ethical issue.

The lines between the first and second approaches will sometimes
blur. It is reasonable to assume that the broad ethical framework of any
society informs the ethics of the healing arts. Hence the historian may
pose some cautious questions of medical ethics that historically informed
common sense suggests as relevant to a period for which no hard evidence that pertains directly to the issue being considered has survived.
For instance, there appears to be no evidence that in the early Middle
Ages, particularly within the context of monastic or sacerdotal medicine,
the issue of active or passive euthanasia arose. Nevertheless, given the
well established emphases in Christian teaching on the sinfulness of both
murder and suicide, the *imago Dei* as the basis for inherent worth of
human life, and the imperative to care for the ill, succor the suffering,
and comfort the dying, it would be patently absurd to suggest either a
positive attitude towards active euthanasia or a conscious refusal to
attempt to do what little could be done to preserve the life of the ill, even
the terminally ill.

The historian should be keenly analytical in every historical study in
which he engages. As he seeks to understand and describe any aspect of
the past, he seriously violates the very integrity of his discipline if he
allows himself to sit in judgment on the past unless he first warns his
audience that he is, for the moment, adopting a role other than an historian's. An economic historian, using modern economic tools of analysis,
must ask questions of the past that are foreign to it in an attempt to
understand better its structure and development. But an historian of
ethics must treat the past differently, taking care lest he impose upon any
ethos concerns, interests, concepts, and definitions foreign to it. Hence he
must exercise considerable restraint lest he seek to wring from the
primary sources answers to questions that would never have arisen in
their original surroundings. In other words, his work should be as free as

possible not only from his own values and preconceptions but also from ethical theories alien to the time he is investigating.

Which of the three approaches mentioned above should one use in attempting to research and write histories of ethics of diagnosis? The third, i.e., attempting to tease primary sources into addressing issues that could not even have been posited in the past but are now titillating suppositions for the philosopher *qua* historian or the historian *qua* philosopher, would be of no value for the historian *qua* historian. The second approach, i.e., attempting to force the sources into addressing questions that could have been raised but apparently were not, can be valuable only if one vigorously refuses to read issues of today into a past environment where they would never have arisen. The first, i.e., addressing only those issues that the sources address, is obviously the safest. It is, however, the least interesting for it enables the historican to discuss only the issues that the sources specifically raise in the chronological order in which they appear. The most profitable and still responsible approach is a combination of numbers one and two that will seek to provide a description of the broad ethical framework within which healers functioned in order to appreciate the ethical considerations that directly or indirectly governed the practice of their art, supplementing this with specific issues when they arise in the literature, and refining the focus considerably as one encounters efforts to create philosophical systems of medical ethics. At every stage of investigation, the historian will be able to say something meaningful about ethics of diagnosis, and to describe or suggest *an* ethics of diagnosis. The historian *qua* historian, however, will be unable to speak meaningfully of *the* ethics of diagnosis.

The first discussion of ethics of some aspects of the diagnostic phase of medical practice does not even constitute the creation or emergence of *an* ethics of diagnosis, much less *the* ethics of diagnosis. I have already asserted that as long as there have been those in any society who have attempted to function as healers, there have been various ethical frameworks at least informing and hence affecting the diagnostic phase of the attempted healing process. Accordingly, there has always been *an* ethics of diagnosis even when no one in a society ever isolated any aspects of that phase of the attempted healing process for ethical consideration. Likewise, any development in the history of nosology that seems so objective from a modern, Western scientific perspective that it makes earlier or contemporary nosologies appear ill-founded, does not constitute the creation or emergence of *an* ethics of diagnosis, much less *the* ethics of diagnosis. This is true even if, in providing a new conceptual framework that significantly altered certain aspects of diagnosis, this nosological development seems to have evinced a new level of ethical sensitivity or to have propounded a new ethical standard.

Perhaps I should clarify what I mean by *an* ethics of diagnosis as distinct from *the* ethics of diagnosis. Many ethical questions may arise in diagnostic processes, and we can, of course, subject aspects of diagnostic processes to ethical scrutiny. When we do so we may think that we are discussing *the* ethics of diagnosis while in reality we are simply considering certain ethical questions involving various aspects of diagnostic processes, *unless*, that is, we stand back from the specifics and seek to raise and examine conceptually the broad ethical framework of that part of medical practice by which diagnostic processes may be informed. Then we must be exceedingly careful lest we simply speak about general ethical considerations germane to medical practice and, by applying them to aspects of diagnostic processes, think that we are saying something about *the* ethics of diagnosis. Even if we avoid the latter and succeed in formulating a conceptual ethical framework for considerations of diagnostic processes, we have yet only created *an* ethics of diagnosis although we may have the temerity to claim that it should be regarded as *the* ethics of diagnosis. Should such a discussion survive the ravages of time, historians would be quite right in dismissing our system that we labelled *the* ethics of diagnosis as simply an attempt at providing a philosophical framework for presenting and defending *an* (i.e. our) ethics of diagnosis. It will have no greater claim to being called *the* ethics of diagnosis than would society's broad ethical parameters under which are subsumed medical ethics in general and perhaps unarticulated ethical principles affecting diagnosis, except that "our" ethics of diagnosis may be more philosophically tidy.

In the history of medical ethics there have not as yet arisen specific systems of ethics of diagnosis. Rather various attempts have been made to formulate philosophical systems of medical ethics that could have provided a philosophical framework for the ethical aspects of any part of the realm of health, sickness and healing. Within such philosophical systems the historian should be able to say something meaningful about ethics of diagnosis for the time and place in question. And in the absence of philosophical systems of medical ethics (in other words, for most of the vast span of history), he must employ the first of the three approaches, supplementing it with the second in order to be able to see at least some aspects of the diagnostic phase of healing practices of the past through the grid of the broader moral perceptions of that society.

What should ethics of diagnosis mean to the historian who is about to explore the history of diagnosis? In a sense, it should mean *nothing* to him *qua* historian. Rather, his concern should be to discover what it meant or (and here he must be very cautious) could have meant it any historical period. The only circumstances under which his own or, for that matter, any other presuppositions concerning ethics of diagnosis can legitimately be brought to bear on his investigations are in preliminary efforts to determine the nature of the questions he should seek to answer and the parameters of the study in which he is to engage.

When the historian seeks to acquire some conceptual orientation for his historical investigation of ethics of diagnosis by fishing in the torrent of contemporary bioethical debate, occasionally casting his line into tributary streams of medical sociology, medical anthropology, and medical economics, he may well land some specimens that are either new species or have evolved unrecognizably from their ancestors. For example, judging from current bioethical literature and sociological analyses, diagnosis appears to be viewed by some as a "warrant to treat". The ethical implications of such a phrase are enormous. Two implications come immediately to mind, one being negative, the other positive. First the positive which is essentially deontological in the etymological sense of deontology, in which "must" or "ought" is inherent: Engelhardt makes very clear some contemporary deontological implications of diagnosis: "Choosing to call a set of phenomena a disease involves a *commitment to medical intervention...*" (my emphasis, [3], p. 137). There are probably few today who would disagree. Indeed many would likely insist that the very act of successful identification of a condition as a disease (or, to be a little broader, an illness) involves an ethical imperative to medical intervention, which may be nothing more than an attempt to treat.

But is this a concept that the historian should bring into his investigations of the history of ethics of diagnosis? Yes, but only in order to suggest the origins and trace the development of this deontological imperative that is at present accepted as inherent in diagnosis, since for much of human history the idea that diagnosis is a "warrant to treat" would have no validity. Indeed the conviction that diagnosis is a "warrant *not* to treat" would have much greater validity, since in many times and places diagnosis of a condition as incurable placed a well accepted ethical imperative on the physician *not* to attempt to treat.

Let us consider the negative implication of the suggestion that diagnosis is a "warrant to treat" insofar as motive for diagnosis suggests itself. I shall extend Engelhardt's statement: "Choosing to call a set of phenomena a disease involves a commitment to medical intervention [and] the *assignment of the sick role ...*" (my emphasis). Few today would argue that when a physician calls a person ill, he does not explicitly assign to that person a sick role or, to put it differently, legitimate the condition or complaint. Eliot Freidson has maintained that

medicine's monopoly includes the right to create illness as an official social role....It is a part of being a profession to be given the official power to define and therefore create the shape of problematic segments of social behavior: the judge determines what is legal and who is guilty, the priest what is holy and who is profane, the physician what is normal and who is sick ([6], p. 37, as quoted by [1], p. 1666).

The ethical burden of such an awesome social (and moral) role as the medical profession exercises in modern, Western society is profound. But is it of value to the historian who is seeking to investigate the history of ethics of diagnosis? Indeed not, for the suggestion that physicians held such enormous social (and moral) power would have been laughable at many (probably most, perhaps all) times in our other-than-recent past. An ethical scrutiny of the motives of physicians as legitimaters of illnesses perhaps would not be inappropriate today, but for the historian to subject the motivations of physicians of the past as supposed *legitimaters of illness* to ethical examination would be exceedingly misleading.

Many other issues that may now be isolated and discussed specificially as areas of ethics of diagnosis, simply because they have emerged as special categories owing to recent socio-medical developments, are either meaningless or misleading for historians except those who are dealing with the very recent past. Since only rarely do the historian's sources focus on specific aspects of the diagnostic phase of medical practice for ethical consideration, he will usually be compelled to consider diagnosis generally and various features of it specifially in the light of broader principles of medical ethics. Hence a history of ethics of diagnosis will invariably say much that will also apply to other areas of medical practice. Or, to state it differently, much, indeed most, of what one can say about ethics of diagnosis consists simply of general principles of medical ethics that apply to all or most phases of medical practice. When any of these features is applied to the diagnostic phase of medical practice, it is not *eo ipso* part of ethics of diagnosis.

Let us consider ethics of diagnosis vis-à-vis ethics of therapy. The *telos* of diagnosis is accuracy. The *telos* of therapy is efficacy. The chief virtues of the physician in respect to the *telé* of diagnosis and of therapy are competence and diligence (the latter being conscientiousness, i.e., suitable thoroughness). Such assessments, incidentally, are not my ahistorical presuppositional tools with which I may mold the evidence to create a history of ethics of diagnosis or of therapy. Throughout history the incompetent physician has been regarded as unethical in attempting to succor his fellows' ills. Hence competence has been and remains an essential feature of the ethical physician. (We shall leave aside the significant implications of the highly relative nature of the idea of competence.) When the historian seeks to apply criteria of competence to the diagnostic phase of medical practice of any time and place, he should be able to say something meaningful about standards of, and reasonable expectations regarding, diagnosis any significant falling short of which would likely have been regarded as unethical. He could, of course, do the same thing in dealing with ethics of therapy. The similarities between what he could say about the criteria of competence in ethics of diagnosis and ethics of therapy of any time and place would probably be significantly greater than the differences. Would the similarities for both be any less part of ethics of diagnosis than the differences peculiar to the diagnostic phase? No; but the differences would, of course, have a uniquely "diagnostic" flavor to them. Any history of ethics of diagnosis that would consider only the differences would be very distorted because of its synthetic compartmentalizing. Any meaningful and historically responsible history of ethics of diagnosis must, when dealing, e.g., with matters of competence, include both the general and the particular. A more likely flaw than compartmentalizing the particular at the expense of the general is to merge both into a supposed ethics of diagnosis, as if the general were the outgrowth of considerations of ethics of diagnosis rather than of general principles of medical ethics that (the historian can suggest) affected diagnostic standards and expectations in various ways. Hence, the history of ethics of diagnosis must be nearly a general history of medical ethics that regularly focuses on ethics of the diagnostic phase but always with reference to the broader scope of medical ethics.

When the historian focuses on the peculiarities of diagnostic processes he must deal with one aspect of the history of medicine that is central to that field, but not of particular importance for the history of ethics of other facets of medical practice than the diagnostic phase: the history of nosology. Then the historian must guard himself from two very subtle pitfalls. One is anachronistically to view nosological systems of the past as more similar to modern, Western models than they actually were. The second can easily result from the first: to regard any such system as providing a better ethical, because more objective, framework for diagnosis than any alternatives.

It is easy for the historian to go awry here. We would do well to heed Owsei Temkin's admonition that

to be historically comprehensive, medicine cannot be defined as a science or the application of any science or sciences. Medicine is healing (and prevention) based on such knowledge as is deemed requisite. Such knowledge may be theological, magic, empirical, rationally speculative, or scientific. The fact that medicine in our days is largely based on science does not make other forms less medical – though it may convince us that they are less effective ([9], p. 16).

One could also say that the fact that medicine in our days is largely based on science does not make other forms ethically inferior.

If the *telos* of diagnosis is accuracy, it follows that ethics of diagnosis involves, first of all, the expectation of accuracy in identifying the patient's complaint. In an historical study, accuracy of diagnosis, however, must not be viewed anachronistically in scientific terms compatible with some current medical model but rather as a condition of consistency of the diagnostician's techniques and conclusions with his culture's overall structure of reality. For instance, illness has typically been an integral part of primitive or archaic cultures' total construct of reality in which the perceived causality of disease is regarded as a vital part of a complex network of purposed contingencies. To diagnose accurately in this context is to diagnose consistently with the structure of such purposed contingencies. The ethical obligations of the diagnostician can, in this environment, be best seen in terms of his consistent function as the interpreter of the considerations of health and illness within the broadest cosmological realities and social values of the community. *Mutatis mutandis*, a somewhat similar situation prevails in a modern scientific setting in which the foremost ethical obligation of the diagnostician is accuracy that is dependent upon both the consistency of his techniques and conclusions that are compatible with that part of society's overall structure of reality, which is the current medical models' categories. Both in primitive or archaic cultures and in modern scientific societies, to attempt to diagnose in any way that is inconsistent with accepted structures of reality is to act unethically, most specifically violating patients' reasonable assumptions of competent presuppositions and, in a broader sense, violating the accepted perceptions of truth or reality that inform those presuppositions. One significant difference between a primitive or archaic milieu and the modern scientific environment is that, in the former, the presuppositions undergirding diagnosis are more obviously integrated into the totality of the culture's ethos, while in the latter the highly technical knowledge of the medical specialist that is often hidden behind a facade of jargon and apparatus isolates him from the patient. As a result, the medical model is more accepted than understood by the laity as being consistent with society's larger scientific perception of reality. It may well be that unethical diagnosis was more easily discernible by the laity in a primitive or archaic culture than it is in a modern scientific society. Furthermore, in spite of the enormous social role of medicine today, definitions and diagnoses of illness in a primitive or archaic culture likely had significantly broader implications because of the social, religious, and moral relationship of people with the complex network of purposed contingencies that was their cosmos. Henry Sigerist observes that "it is an insult to the medicine man to call him the ancestor of the modern physician. He is that, to be sure, but he is much more, namely the ancestor of most of our professions" ([8], p. 161). It is only with the increasing medicalization of Western society that the physician's social role is even beginning to approach the importance that the medicine man has typically enjoyed.

When studying the history of medicine, we can so easily be beguiled into thinking that there is something remarkably scientific and hence refreshingly objective in those nosological systems that are ontological rather than physiological. Ontological nosologies tend to view diseases as species and each disease as an entity. From Paracelsus' idea of disease as a parasite and Harvey's theories of contagion through Sydenham's and Bichat's *historiae morborum* and Koch's postulates to modern bacteriology, ontologists have been repeatedly accused of wrongly melding into one unreality *causa morbi* and *ens morbi*.

Those who accuse ontologists of this error are the physiological nosologists who are, interestingly enough, in the tradition of the founders of what is often mislabelled scientific medicine, that is, Greek medicine. The victory of Greek medical theory was not only the victory of "rational" constructs over "irrational" or "superstitious" concepts of disease but was also the victory of physiological nosology over the most consistent ontological nosology in the history of medicine, that is primitive or

archaic demonic disease etiology in which every *ens morbi* is its own
causa morbi, a nosological system that viewed disease wholly etiologi-
cally rather than symptomatically.

Influenced, as they were, by Greek natural philosophy that insisted on
defining and explaining reality through naturalistic rather than
mythopoeic constructs, "Hippocratic" physicians likewise insisted on
defining sickness in terms of natural rather than supernatural proximate
causality. Were these physicians, who used their new constructs of
reality, ethically superior, in their diagnosis, to the older, non-rationalistic
healers whom they, for the most part, gradually superseded? Who is ethically superior in his diagnosis, the modern, Western physician who
proves very successful in improving medical treatment in some third-
world countries, or the indiginous medicine man, who is sometimes
amazingly effective when confronted with those states in the face of
which Western medicine and its constructs appear so helpless?

We would do well, when considering ethics of diagnosis of both the
past and the present, to ponder the implications of Eric Cassell's opinion
that it is

> more important that the doctor's explanations be culturally consonant than that they be
> true, since all explanations of illness as well as role relationships between doctor (healer)
> and patient are related to the belief system of the culture in which they occur.... This is
> underscored by the fact that explanations of illness for most of the world's history and in
> most cultures have proved inadequate or false with the passage of time and the progress of
> scientific inquiry, despite the fact that they served their function while extant ([2],
> p. 1673).'

How, then, do we assess the ethical dimensions of changing medical
models? There are times when the historian could find himself regarding
progressive and perhaps more competent physicians of the past as less
ethical in diagnosis and conservative and perhaps less competent physicians of the same time as more ethical.

Are or were physicians who refuse or fail to diagnose in accordance
with accepted medical models incompetent or progressive, ethical or
unethical? Many people, both physicians and laymen, would regard as
unethical a physician practicing after Koch's time who refuses or fails to
use the laboratory sciences as diagnostic tools. But what if the physician
insists on seeing germ theory or its various developments as accountable
for *all* conditions deserving treatment because he holds firmly to a
dichotomy between organic and functional conditions, regarding the
latter as either suspect or outside the physician's purview? In this regard

it is interesting to note that in Britian the General Medical Council, which establishes and enforces "standards of medical education and medical ethics in the interests of the protection of patients", may censure members for breaches of medical etiquette, but "does not claim to be able to remove a doctor's name from the Register 'on account of his adopting or refraining from adopting the practice of any particular theory of medicine'" ([10], pp. 987–988).

When one's physician exhausts pathoanatomical and pathophysiological avenues of diagnosis and pursues a pathopsychological mode of inquiry, the patient may well become uncomfortable and feel that the physician by such probing is acting unethically. All the more so may the patient in frustration question the ethics/competence of the physician who bypasses pathoanatomical and pathophysiological diagnostic efforts and suggests a pathopsychological explanation when the patient, for example, simply wanted the doctor to order a CBC that would reveal whether or not he was anemic.

The historian who happens at times to be a patient may perhaps find himself in such positions but his reaction to ethical questions in these personal contexts should not create or strengthen biases that would affect his analysis of the history of ethics of diagnosis. They may, however, make him aware that as an historian he can too easily focus on physicians' ethical pronouncements, codes of professional ethics, and philosophical systems of medical ethics and ignore patients' views that have been articulated throughout the centuries. He should consider the significance of Cassell's observation that

the doctor's rational system of explanations reinforces the patient's system of reason so that the mysteries of illness are again contained. Reflect, in this aspect, how important it is that the patient's illness have a name, since the word is often seen as containing the thing.... The disease concepts of Western medicine can be seen in this light as a highly useful and effective conceptual structure ([2] p. 1673).

But the historian of ethics of diagnosis should be sensitive to the extent to which people may disagree with such assessements. For not every patient would regard as an ethical imperative the physician's sense of duty to label, as Matthew Arnold's lines so poignantly show:

> Spare me the whispering, crowded room,
> The friends who come, and gape, and go;
> The ceremonious air of gloom –
> All, which makes death a hideous show!

> Nor bring, to see me cease to live,
> Some doctor full of phrase and fame,
> To shake his sapient head, and give
> The ill he cannot cure a name.

As an ethical issue it is debatable whether the perceived necessity of giving an illness a name may well not meet a greater need of the physician than of the patient (or his next of kin). Likewise the historian of ethics of diagnosis should consider not only physicians' but also patients' views and needs when dealing historically with the ethical dimension of that hoary problem of "what to tell the patient" or "truth-telling in medicine".

The task of the historian of ethics of diagnosis is molded by positive and negative strictures. On the one hand, he needs to be very sensitive 1) to the significant degree of fluctuation there has been in the number and variety of issues subsumable under "ethics of diagnosis" and 2) to the enormously varying importance of diverse and changeable social factors over time. On the other hand, he must beware lest he impose upon the past 1) any ethical grid historically alien to it, 2) the unprecedented variety of issues that today should be considered under the rubric "ethics of diagnosis", and 3) the increasingly complex and unique social matrix of modern institutionalized, bureaucratized "health-care delivery systems".

Western Washington University
Bellingham, Washington, U.S.A.

NOTE

* I wish to thank Dr. Gary B. Ferngren for his astute criticism of a draft of this paper and for transforming my sometimes turgid and ponderous prose into something more palatable.

BIBLIOGRAPHY

1. Bloom, S.W.: 1978, 'Therapeutic Relationship: Socio-historical Perspectives', in [7], pp. 1663–1668.
2 Cassell, E.: 1978, 'Therapeutic Relationship: Contemporary Medical Perspective', in [7], pp. 1672–1676.
3. Engelhardt, H. T., Jr.: 1975, 'The Concepts of Health and Disease', in [5], pp. 125–141.

4. Engelhardt, H. T., Jr.: 1986, *The Foundations of Bioethics*, Oxford, New York.
5. Engelhardt, H. T., Jr., and Spicker, S. F.: (eds.), 1975, *Evaluation and Explanation in the Biomedical Sciences*, D. Reidel, Dordrecht, Holland.
6. Freidson, E.: 1970, *Profession of Medicine: A Study of the Sociology of Applied Knowledge*, Dodd, Mead and Co., New York.
7. Reich, W. T. (ed.): 1978, *The Encyclopedia of Bioethics*, Free Press, New York.
8. Sigerist, H.: 1951, *A History of Medicine*, Vol. 1, Oxford, New York.
9. Temkin, O.: 1977, *The Double Face of Janus*, Johns Hopkins, Baltimore.
10. Wolstenholme, G.: 1978, 'Medical Ethics, History of: Britain in the Twentieth Century', in [7], pp. 987–992.

H. TRISTRAM ENGELHARDT, JR.

OBSERVER BIAS: THE EMERGENCE
OF THE ETHICS OF DIAGNOSIS

One of the difficulties in writing history is that ideas change through
time. When we describe what took place in the past, we do it through the
ideas of the present. But the past had its own ideas, its own priorities, its
own ways of seeing what mattered. Generally, we are more at home with
recognizing this circumstance when we look at the history of religion or
politics. We as a culture feel more comfortable in acknowledging that
religious and political ideas change over time. We tend to look at science
and knowledge in more absolute terms. After all, science tells us what is,
and what is simply is. This view has only recently been seriously challenged. Over the last two hundred years or so we have begun to take
history seriously, and the possibility that the very appreciation of reality
changes through time. In part, this has been due to the reflections of
Giambattista Vico (1668–1774) and G.W.F. Hegel (1790–1831). Even
the idea of scientific revolutions can be traced back to at least Hegel's
Philosophy of Nature, where he says, "All revolutions, whether in the
sciences or world history, occur merely because spirit has changed its
categories in order to understand and examine what belongs to it, in
order to possess and grasp itself in a true, deeper, more intimate and
unified manner" ([13], Sec. 246 Zusatz, V of I. p 202). Hegel recognizes
that ideas and reality are mutually implicative.

Much has been made recently of the fact that what we mean by
science changes through time and that our appreciations of reality change
with those changes. This point was developed with examples drawn from
medicine (changes in the concept of syphilis in particular) in the 1930s
by Ludwik Fleck through a book with the provocative title, *The Genesis
and Development of a Scientific Fact* [5]. The idea of scientific revolutions was then popularized by Thomas Kuhn [9]. From the 1960s through
the 1980s there has been, despite many qualifications and criticisms [10],
a recognition that developments in science are not simply incremental,
but are tied to fundamental shifts in what we mean by knowing and
doing science.

The history of diagnosis provided in this volume by Laín-Entralgo,
Diego Gracia, Albarracin, Peset, and Amundsen show the extent to

63

J. L. Peset and D. Gracia (eds.), The Ethics of Diagnosis, 63–71.
© 1992 *by Kluwer Academic Publishers. Printed in the Netherlands.*

which knowledge in general and medical knowledge in particular have been understood to be problematic and to involve matters of judgment with ethical implications. As Entralgo demonstrates, the ancients appreciated that knowing well was a difficult task. Hippocrates rejects empty postulates (*Ancient Medicine I*) and emphasizes that theory must be founded on facts and the actual observation of phenomena (*Precepts I*). However, when it comes to describing diseases in the *Epidemics*, for example, the author does not communicate any recognition of the circumstance that his presuppositions may be restructuring the facts he sees. Nor in his *Regimen of Acute Diseases* is there skepticism about his capacities to treat. As a consequence, it is very difficult for us to decide what diseases he is describing because his and our views of relevant findings are quite different. Though there were skeptics in the ancient world, and though there was appreciation of the difficulty of knowing truly, there was little of our historically articulated realization that in knowing we as much create as discover reality. As a result, Laín-Entralgo properly notes that the ethical difficulties that confronted the ancient diagnostician were associated with knowing truly what was present. The problem was that of making sure that claims coincided with objective reality.

Much the same can be said about the Middle Ages with the qualification Diego Gracia introduces: treatment was then seen within a well-developed religious moral viewpoint. There was little doubt about the truth of that framework. Ethics shared with medicine a cultural assumption that one could discover that which is, or should be, the case. The modern world may have initially secularized ethics, as Albarracin shows, but it also led to a new religion of medical facts and the hope that what could be objectively disclosed would be objectively normative. As one enters the latter part of the eighteenth century and finally the nineteenth, as both Albarracin and Peset correctly note, social concerns emerge. On the one hand, codes of medical ethics are written to control and coordinate medical practice. On the other hand, societies and nations begin to respond to meet perceived medical needs through organized systems of health care so that the provision of health care becomes part of the morality of good public policy. In all of this, medicine is seen ever more in terms of general ethical and moral considerations. Diagnosis is no longer merely the act of an individual physician attempting to know truly.

The major development is a refined skepticism regarding the capacity to know truly and to intervene effectively. As Peset points out, this was appreciated by both Josef Skoda (1805–1881) and Thomas Addison (1793–1860). Therapeutic nihilism provides only a partial indication of the general change in the understanding of what it is to know truly that fashioned our contemporary appreciation of medical knowledge. Our current critical understanding of medical knowledge is the result of a partial loss of faith in ourselves as knowers. Consider, for contrast, how Soranus (A.D. 98–138) presented treatment options. In his work on gynecology he lists remedies with very little embarrassment about their heterogeneity and number. For example, he introduces a discussion of contraception by advising that "during the sexual act, at the critical moment of coitus when the man is about to discharge the seed, the woman must hold her breath and draw herself away a little, so that the seed may not be hurled too deep into the cavity of the uterus. And getting up immediately and squatting down, she should induce sneezing and carefully wipe the vagina all round; she might even drink something cold" ([15], p. 64). After giving a list of other approaches based on extracted semen, Soranus lists a large number of other means of contraception, including those that can draw semen from the uterus: "Moreover to some people it seems advisable: Once during the month to drink Cyrenaic balm to the amount of a chick-pea in two cyaths of water for the purpose of including menstruation. Or: Of panax balm and Cyrenaic balm and rue seed, of each two obols, [grind] and coat with wax and give to swallow; then follow with a drink of diluted wine or let it be drunk in diluted wine" ([15], p. 65). One has a list of approaches with little systematic theoretical account of why they should succeed or why one should prefer one over others.

Theoretical accounts in terms of humoral and other doctrines during ancient and medieval times did not improve matters. Consider, for example, an account of hydrops given by John of Gaddesden (1280–1361). "Hydrops is a material sickness of which the cause is a cold matter, overflowing and entering into the limbs, and thence arise all its manifestations,... Avicenna says 'Hydrops is an error of the combining energy (virtutis unitivae) in the whole of the body, following on a change of the digestive energy in the liver'" ([8], p. 82). In terms of this theoretical account, which is indebted directly and indirectly to ancient authors, John of Gaddesden could then forward recommendations for treatment. "The cure of hydrops is of two kinds, common and proper. The proper is by means of various appropriate medicines and by local measures. The common, as says Avicenna, is by extraction of the watery humidity and its drying up, and this extraction may be carried out in four ways, as Constantine lays down in the seventh book of his Practice" ([8], p. 82). This account of disease is marked not only by appeals to authority but by its robust confidence in reason's capacity to discern structure in reality. Greek and medieval authors such as Aristotle and St. Thomas Aquinas presumed a consanguinity between mind and reality: The agent intellect could abstract an intelligible structure from reality. This congeniality of the knower and the known was fortified by the Christian world view, for reality and the knower were both created by the same God. Scholastic approaches to medicine and to theology could proceed with a firm basis for trust in reason's capacity to disclose reality.

With the beginning of modern science and medicine, a skepticism develops, in part as a reaction against the overly rationalistic, insufficiently empirical approaches of Scholastic philosophy. This recognition of the difficulty of knowing truly is reflected in the writings of Sir Francis Bacon and, through his influence on Sydenham, in medicine. Thomas Sydenham (1624–1689) acknowledged that the confusion of remedies was an embarrassment to medicine. "We are over-whelmed as it is, with an infinite abundance of vaunted medicaments..." ([14], Sec. 17, p. 18). Sydenham hoped to rectify this difficulty by developing a *methodus medendi,* which could be established on the basis of "a long continuance and a frequent repetition of ... experiments ..." ([14], Sec. 16, p. 17). The truth had to be dragged out of reality by encountering the empirical. Though his actual practice of medicine often involved an uncritical acceptance of conventional cures, from a theoretical point of view at least, Sydenham has a modern ring because of his concern about how to establish the efficacy of treatments.

Sydenham suggested that we make a step toward securing medical knowledge by abandoning theoretical speculations about underlying causes of disease. Here he applied the mixture of skepticism and faith in reason that he derived from Bacon. He was concerned that theoretical assumptions would cause us to see reality not as it is, but as we would want to see it. "No man can state the errors that have been occasioned by these physiological hypotheses. Writers, whose minds have taken a false colour under their influence, have saddled diseases with phenomena which existed in their own brains only..." ([14], Sec 9, p. 14). This is a very modern preoccupation. However, he believed that we can recognize our biases, remove them from our minds, and see reality as it is. In this regard, he shared much with the ancients. He assumed a congeniality between the knower and the known. He presumed a capacity to see reality and describe its structures as it is presented to us. As a result, Sydenham recommended that "In writing the history of a disease, every philosophical hypothesis whatsoever, that has previously occupied the mind of the author, should lie in abeyance." ([14], Sec. 9. p. 14) With that discipline one would then be able to describe diseases and one would find that they fall into clear and constant species. "Nature, in the production of disease, is uniform and consistent; so much so, that for the same disease in different persons the symptoms are for the most part the same; and the selfsame phenomena that you would observe in the sickness of a Socrates you would observe in the sickness of a simpleton" ([14], Sec. 12, p. 15). The mind is able to reach out and know reality. Moreover, the primary reality for the physician is clinical reality.

Sydenham's views fit well within the assumptions of the naturalistic method. Naturalists assumed that one could observe reality and, if one were careful to avoid prejudices, come to know it. Care in inspecting nature would be enough. Though physicians had begun to recognize the difficulties of observer bias, they had not suspected the depth or the range of that bias. There was none of our modern concern to experiment systematically, to manipulate nature systematically.

Before a sense of systematic experimentation in medicine was fully developed, another major epistemological shift would occur in medicine. Describing clinical findings in a clinical fashion would no longer be seen to be sufficient. Instead, one would look to what could be revealed by postmortem anatomical dissections, cellular pathology, physiology, and later microbiology. Initially it was simply as if the naturalists had turned inward to describe foundational anatomic structures. An in many ways helpful description of this shift in epistemological attention is given by Michel Foucault in his *Birth of the Clinic* ([6], [7]). However, the title and the work do not sufficiently acknowledge that, instead of the birth of the clinic, one saw the subordination of the clinic to the laboratory [4]. Clinical medicine was no longer the basic science. The basic sciences were now anatomy, physiology, and microbiology.

The result was two tiers of medical reality, one clinical, the other laboratory. Each could propose questions for the other, and in principle correct the other, though the sphere of laboratory medicine was considered to be primary or basic. A critical dialectic was initiated between the clinician and the laboratory scientist. Laboratory scientists could demand that clinicians have their descriptions of diseases conform to what was known about the basic processes of pathology. Clinicians, on the other hand, could challenge laboratory scientists to provide convincing accounts of what was seen in everyday practice. Each domain of objectivity, each domain of encounter with reality, could strengthen and criticize the other. This interplay between domains of medical reality heralded the fact that the days of the naturalists in medicine were over. One could no longer take for granted the significance of what one saw. Medical reality had to be understood and assessed within a complex of interlocking observations, explanations, and theories.

But even this was not enough. In the beginning of the nineteenth century, individuals such as Louis and Gavaret introduced the need for statistical methods in order to diminish observer bias ([7], [11]). It was Louis who showed through a statistical study that bleeding did not contribute to the treatment of inflammatory diseases. Finally, in the twentieth century this appreciation of the difficulty of knowing truly would result not only in a practice of carefully controlled and repeated experiments, but also in the use of double-blind randomized studies. The best-intentioned observer could observe falsely. Only vigorous compensations for observer bias could produce objective results, even if one no longer regarded objectivity as correspondence with the object. Even to gain objectivity as intersubjectivity, one had carefully to discipline the process of knowing.

The development of the *methodus medendi* sought after by Sydenham involved major communal undertakings in order to determine whether particular treatments were safe and effective. Large-scale trials had to be designed and undertaken by various investigators, aided not only by better theoretical understandings of disease but statistical means of compensating for observer bias. The ethics of diagnosis came to incorporate, at the very minimum, an obligation to overcome observer bias.

The foregoing accent only the moral problems ingredient in knowing truly. Medicine, however, is not undertaken to know the world truly, but rather to intervene effectively. Medicine is not a pure but an applied science. It is with this consideration that another set of ethical issues emerges. How does one determine whether a medical intervention will do more good than harm, especially since patients may not endorse the same hierarchy of costs and benefits as do physicians. The acquisition of knowledge involves not only financial costs, but exposures of patients to risks of morbidity and death. In addition, diagnoses are rarely certain or even close to certain, so that in choosing among different diagnoses one runs the risk of exposing patients to over-treatment or under-treatment. In order to manage these risks, special clinical classifications have been created, such as those that stage cancers in order better to coordinate treatment in terms of the available information. The goal in such undertakings is to array actions with likely states of affairs. But, different clinical approaches to characterizing reality have different possible costs and benefits. The question then arises of who should set what weights to what values for which outcomes.

At this point, the ethical issues associated with diagnoses become sailent. It had been clear for a long period of time that there are ethical issues involved in deciding when to treat. One might think here, for example, of the development of the concept of ordinary versus extraordinary care. Within Roman Catholic moral theology, it was recognized that significant moral issues are involved in deciding when a treatment carries with it an "undue" burden for the patient [3]. What is new is a clearer recognition that diagnoses are treatment warrants and that moral judgments are integral to an applied science where one is asked to intervene, usually with less than conclusive evidence, in ways that have significant implications for the suffering, well-being, and life of a patient [16]. It is not simply that knowledge is uncertain, but that it is uncertain how the knowledge should be applied. Choices among different ways of describing facts and different ways of evaluating outcomes are central to the clinician's predicament. Here the issue is not simply observer bias, but the bias of the intervener, the clinician. This issue is central to current debates in bioethics. Again what is new here is a better appreciation of the circumstance that medical knowledge is not only theory-infected, but value-infected. The nineteenth century produced a series of reflections on the character of medical knowledge [1,2,12]. There was not, though, an appreciation of how value priorities are reflected in medical descriptions of reality.

The recent development of interest in bioethics is a recognition of the difficulties involved in medical interventions and that choices of treatment are not objective choices in any straightforward, value-free sense of the term. To talk of the ethics of diagnosis is to recognize the intimate interplay among explanation, evaluation, and intervention. A serious, sustained examination of this enterprise is new in the history of medicine and reflects a critical appreciation of the difficulties involved in medically knowing and intervening well. It is new because it requires recognizing that diagnoses are not merely claims to knowledge. They are claims to knowledge within an intervention-oriented applied science, so that a diagnosis as a therapy warrant conveys with it presumptions regarding the reasonableness of particular therapeutic interventions. One must be careful in making diagnoses because of their implications for treatment. One must be very cautious in making a diagnosis that will imply that a risky treatment is indicated; one need be less concerned if the treatment indicated by a diagnosis carries with it little risk of any serious side effects. The ethics of diagnosis involves recognizing not only the theory-laden but the value-laden character of medical knowledge. As a result, one attempts to describe truly in order to intervene effectively in the sense of realizing certain therapeutic goals. In addition, one must shape the structures of medical reality well in order successfully to direct physicians to effective treatment. Because the good of patients is at stake, the construal of reality must take place with moral issues in mind. And the character and the process of the construal have moral implications. Though this understanding of medical knowledge is a relatively recent one, it is philosophically important, for it opens up a wide range of issues regarding how to assess the interplay for truth and the commitment to achieve what is good. Discussions of the problems involved in computer diagnosis in the penultimate section of this volume give one illustration of the questions occasioned by this understanding of medical knowledge.

Baylor College of Medicine
Houston, Texas U.S.A.

BIBLIOGRAPHY

1. Bieganski, W.: 1894, *Logika Medyzyny*, Kowalewski, Warazawa.
2. Blane, G.: 1819, *Elements of Medical Logick*, T. and G. Underwood, London.
3. Cronin, D.: 1958, *The Moral Law in Regard to the Ordinary and Extraordinary Means of Conserving Life,* dissertation submitted to Gregorian Pontifical University.
4. Engelhardt, H.T., Jr.: 1986, *The Foundations of Bioethics*, Oxford, New York.
5. Fleck, L.: 1935, *Entstehung und Entwicklung einer wissenschaftlichen Tatsache*, Benno Schwabe, Basel; 1979, *Genesis and Development of a Scientific Fact*, ed. T. Trenn & R. Merton, trans. F. Bradley & T. Trenn, University of Chicago Press, Chicago.
6. Foucault, M.: 1963, *Naissance de la Clinique*, Presses Universitaires de France, Paris.
7. Gavaret, J.: 1840, *Principes generaux de statistique medicale*, Bechet jeune et Labe, Paris.
8. John of Gaddesden: 1942, 'Rosa Anglica practica medicine a capite ad pedes', in L. Clendening (ed.), *Source Book of Medical History*, Dover, New York.

9. Kuhn, T. S.: 1970, *The Structure of Scientific Revolutions*, 2nd ed., University of Chicago Press, Chicago.
10. Lakatos, I. and Musgrave, A. (eds.): 1970, *Criticism and the Growth of Knowledge*, University Press, Cambridge.
11. Louis, P.C.A.: 1835, *Recherches sur les effets de la saignee dans quelques maladis inflammatoires*, Bailliere, Paris.
12. Osterlen, F.: 1855, *Medical Logic*, ed, and trans. G. Whitney, Sydenham Society, London.
13. Petry, M.F. (ed. & trans.): 1970, *Hegel's Philosophy of Nature*, Humanities Press, New York.
14. Sydenham, T.: 1979, *Medical Observations Concerning the History and the Cure of Acute Diseases*, in *The Works of Thomas Sydenham*, trans. R.G. Latham, Classics of Medicine Library, Birmingham, Alabama.
15. Temkin, O. (trans.): 1956, *Soranus' Gynecology*, Johns Hopkins Press, Baltimore.
16. Wulff, Henrik: 1976, *Rational Diagnosis and Treatment*, Blackwell Scientific, Oxford.

SECTION II

ANTHROPOLOGICAL INTERPRETATIONS

H. TRISTRAM ENGELHARDT, JR.

THE BODY AS A FIELD OF MEANING:
IMPLICATIONS FOR THE ETHICS OF DIAGNOSIS

Evaluation, explanation, and the ethics of intervention are tightly intertwined in medicine, because medicine intrudes intimately into the minds,
bodies, and lives of humans. As Jose Mainetti [2], Stuart Spicker [3], and
Thomas Bole [1] show, diagnoses are evaluative because concepts of
disease, illness, affliction, and deformity presuppose judgements about
when the minds and bodies of humans fall short of physiological or psychological ideals. To identify something as a disease or illness is to judge
that it is a state of affairs that fails to realize some view of how human
bodies and minds ought to be. Circumstances are regarded as diseases or
deformities rather than as interesting psychological, anatomical, or physiological variations, because they involve a pain or discomfort worthy of
remedy, a lack of grace or form worthy of treatment, or a loss of a usual
or generally desired human function. One invests labor in making a diagnosis not simply in order to know truly, but because one would hope to
be able to avoid or mitigate some unpleasant state of affairs. In the case
of prognosis, one wants at least to be able to plan for likely unpleasant
future developments.

The rationale that moves a clinician to expose a patient to pain, discomfort, or even a risk of death while collecting information to make a
diagnosis, or that establishes certain treatments as indicated for particular
diagnoses, presupposes a view about what limitations, pains, or dysfunctions merit intervention, when and how. If one were to consider what was
involved in writing a computer program for medical diagnosis, one
would need to take into account not only how to establish a correct diagnosis, but how to establish it with as few costs in patient morbidity and
mortality as possible. The logic of medical diagnosis would need to take
into account how patients weigh or regard different intrusions undertaken
in collecting information in order to make a diagnosis. It would need to
take into consideration how patients are likely to regard likely side
effects of treatment.

Many of the judgements patients make about the reasonableness of
diagnostic and therapeutic interventions depend on their views of their
bodies and the significance for them of their bodily functions. Human

J. L. Peset and D. Gracia (eds.), The Ethics of Diagnosis, 75–77.
© 1992 *by Kluwer Academic Publishers. Printed in the Netherlands.*

bodies are, after all, not mere objects, as moral judgements concerning their desecration show. Moreover, the moral outrage associated with mutilation of a corpse depends also on the body parts removed, suggesting that not all parts of our bodies have the same significance. Thus, our faces, which present us to the world, tend to have greater significance than our toes. One might think of the revulsion in the latter part of the Byzantine Empire associated with the loss of the nose, which led to excision of the nose being recognized as a severe punishment. Indeed, the loss of a nose was generally an impediment to becoming emperor, an impediment that was set aside through the use of a gold prosthesis in the case of Justinian II.

Our bodies realize us in the world so that limitations on our bodies are limitations on us. Medicine's engagement in treating our bodies is thus directed by the range of our concerns about our bodies. The enlistment of cosmetic surgery both to restore features destroyed by trauma (e.g., severe burns) as well as to realize particular visions of anatomical grace springs from a fundamental concern with how we are presented to others and regarded. Diseases or defects that blunt our capacity to see, hear, feel, smell, and taste cut us off in very radical ways from the world. What is at stake ranges from loss of elements of the aesthetic to a truncation of our capacity to determine and appreciate the range of reality. On the other hand, diseases that limit the use of our limbs or even involve complete paralysis threaten to limit our capacity to be agents and engage the world and others. Diseases such as amyelotropic lateral sclerosis deprive their victims of being an instance of *homo faber* and reduce them to a contemplative life of sense reception without response. Other diseases limit self control. They range from tremors and choreas to loss of anal and urinary sphincter control. One is no longer able to deport oneself with grace. One is constantly threatened with embarrassment. One's own body in its failure to respond to the will becomes a threatening other. Other diseases alter special senses we have of ourselves and special avenues of social engagement. Cancer of the breast or of the vagina and diseases such as diabetes or cancer of the prostate, which can lead to impotence, can alter one's self-image and one's life with others. In these, as in many other areas, limitations of function and of anatomy provide the occasion for medical treatment.

Baylor College of Medicine
Houston, Texas, USA

BIBLIOGRAPHY

1. Bole, T.J.: 1992, 'Anthropology and the Hidden Values in Diagnosis', in this volume, pp. 123–127.
2. Mainetti, J.A.: 1992, 'Embodiment, Pathology, and Diagnosis', in this volume, pp. 79–93.
3. Spicker, S.F.: 1992, 'Bodily Integrity, Trust-Telling and the God Physician', in this volume, pp. 107–122.

JOSÉ ALBERTO MAINETTI

EMBODIMENT, PATHOLOGY, AND DIAGNOSIS

I. INTRODUCTION

Diagnosis is not knowledge for knowledge's sake. It is knowledge for the sake of action. Medicine exists in order to cure, to care, to intervene, or, in limiting cases, to know when not to intervene. Medicine is not a contemplative science. Treatment carries with it presuppositions, views about when our bodies are whole and healthy and when our bodies require the intrusions of medicine. One cannot understand the wide range of medical actions without first analyzing how we view our embodiment as successful or unsuccessful, complete or incomplete, flawed or well accomplished. Medicine is not a science that attempts to see man sub specie aeternitatis; it regards man sub specie pathologicae. It is man as sick, man as pathological, who is the object of medical attention. Medicine carries with it a very special anthropology, a study of the logos of the human, not directly in terms of human ideals, but rather first and foremost in terms of human shortcomings. Despite the World Health Organization, health is generally understood in terms of disease, in terms of a wide range of shortcomings, limitations, and failings. The positive, the truly human is a negation of a negation, a negation of the shortcomings that are a negation of particular human goals.

Medical interventions are fraught with ethical consequences and value presuppositions. To see a particular medical treatment as indicated or proper is to understand it in terms of an anthropology of the pathological, which makes certain interventions plausible or proper. Since the pathological is understood in terms of suffering, and suffering is appreciated in terms of human ideals, all interventions take place against a rich background of evaluations. Since all may not share in the same views regarding which limitations should be set aside and how, and since the cooperation of physician and patient requires agreements concerning values and understandings about what is important in being human, serious ethical issues are at stake. To treat while taking proper ethical regard of patients requires acknowledging the range of values embodied in the bodies of humans.

79

J. L. Peset and D. Gracia (eds.), The Ethics of Diagnosis, 79–93.
© *1992 by Kluwer Academic Publishers. Printed in the Netherlands.*

This essay is offered as a step toward making clear the importance of the values ingredient in our understanding of pathology. I will show that one cannot talk about the ethics of diagnosis without taking seriously this background of values. First, I will sketch how concepts of finitude interplay with notions of sickness and illness. Medicine reflects a wide range of cultural appreciations of human imperfections. Then I will make some suggestions about what it means to talk of an anthopology of the pathological. Medicine offers a rich field for philosophical anthropology, which is significant in its own right beyond its implications for helping us to appreciate medicine and its technologies better. I will conclude by showing why an anthropology of the pathological requires acknowledging the patients' experience of suffering. It is suffering as experienced by patients that is the usual warrant for medical interventions.

II. SICKNESS AND FINITUDE

Man is a limited being who is conscious of his limitations and whose actions form a permanent and always renewed attempt to overcome them. To give an example, man, and as far as we know only man, is conscious of his radical limitation, death, and at the same time aims at overcoming this limitation and at the absolute and infinite. The awareness of his fugacity and the simultaneous desire for infinitude cause man to be eternally dissatisfied with this condition, a "tragic", "metaphysical", "supernatural" animal. It is not surprising, then, that man, the human existence, appears to us like something incomplete, restless, wretched, and basically "sick". The philosophical anthropology of today (the philosophy of existentialism, anthropobiology) has stated precisely and sufficiently this characteristic of man, the "infirmitas", a constant in the history of anthropological thought, based on which we can then assume a reflection on sickness as *anthropinon* [14], an introduction to the fundamental belief in the relationship between the problem of embodiment and pathological anthropology.

We will give a brief review of the history of pathological anthropina in order to clarify this concept and then explore the possibility of founding a system of anthropopathology upon it. It might be interesting to trace the history of anthropological thinking in the light of the idea of man as a "sick being" which appears in various meanings and nuances since mythical anthropology up to the anthropological reflections of today. Maybe the origin of the "pathological" definition of the specifically human element can be found in the inherent "abnormality" of anthropology.

For man has always understood himself in comparison with what he himself is not, such as an animal (theriomorphism) or a god (theomorphism). There exists a whole line of "negative anthropology" (similar to a "negative theology") and an "anthropological pessimism" which can be dated back to mythical thinking and its two currents in our culture, Greek "naturalism" (Prometheus) and Judeo-Christian "spiritualism" (Adam). The "gift of fire" that Prometheus procures for men compensates for a biological flaw that condemns them to extinction. Likewise, the "biblical fall" represents an *infirmitas* or loss of the original condition, which opens the way for grace or redemption. Already one sees man as an "erect ape" (devitalized) or as a "fallen angel" (disgraced) that step from the natural to the supernatural and vice versa, which is pointed out through the contranatural, physical, or moral sickness (sin). The ambivalence of the pathological element, negative and positive at the same time, its defective aspect that generates a perfective sense, constitutes the topic of a certain romantic style of anthropology open to the idea of man as "sick animal", which can be traced back to a number of thinkers of varied backgrounds (e.g., St. Augustine, Pascal, Kant, Herder, Schiller, Schopenhauer, Nietzsche). "An anthropoid ape" – writes Unamuno, summing up this idea correctly – "once had a sick son, sick from a strictly animalistic or zoological viewpoint, truly sick, and this sickness besides being a weakness turned into an advantage in the struggle for perseverance" ([33], p. 23). But not always has the equilibrium in the anthropopathological concept been maintained and from sickness as ferment of culture one has gone to sickness as a corrosive of life, to man as "depraved animal" [27] or even as irredeemably sick and pathogenetic, and thus in favour of an irrational vitalism ([10], [15]), according to which the spirit (*Geist*) appears as the "enemy of the soul" (*Widersacher der Seele*) or even as main agent in the destruction of life. Together with Nietzsche's idea of Superman, there rises the idea of an irreversible biological decadence caused by the "noxious bacillus of the mind" ([10], p. 67).

It is evident that the definition of man as a "sick being" is an analogous use, a metaphorical broad sense, of the concept of sickness. It requires further argument even when the transposition of a primarily biological category to a metaphysical level has many times served to underline the ontological negativism of man, the existential absurdity, the error of nature, the imperfectly created being, the useless passion, the radical need as the substance of our life, and even "the sin of being born" ([1], p. 6). What seems to be true is that human nature reveals a structure that we can call *infirmitas, infirma species* [31], *labilitas* [4] or *enfermabilidad* [tendency to become sick] [12]. The Augustinian "uneasiness" and Heideggerian "anxiety" indicate this metaphysical character of human reality, precisely what Lain Entralgo calls "enfermabilidad", this sick modality of our condition through which we really become sick. "The enfermabilidad is an essential part of being healthy: being healthy man cannot help but being sick. Health – a pessimist once said – is a transitory state which leads nowhere good" ([12], p. 19). We must not, therefore, consider human nature as sick but man's condition or "enfermabilidad", the inability not to be sick, which thus is the condition of the possibility of a concrete sickness. In that sense, and only in that sense, can one talk about "pathological anthropina", anthropina meaning according to Landman "the unchangingly fixed 'timeless' basic structures of human existence" ([14], p. 45), the categories of existence or "existentials" of Heidegger. The task of anthropological philosophy is to establish a system of anthropina, distinguishing between fundamental and derived structures. The task of philosophical-medical anthropology is to consider the pathological anthropina among the former, as transcendental basis for all anthropina. And it is just that which we intend when we describe sickness or "enfermabilidad" as a "physical imperfection", "consciousness of alterity", and "conflict of limitations".

A. *Sickness and Biological Imperfection*

A point of departure in anthropological reflection is the old observation that man, compared with animal, seems to be a poorly endowed, helpless, unfinished being with a constitutional deficiency such that he does not have a right to survive and much less a right to impose himself on the world of nature, dominating it at will. In contrast with the beast, the human element is precisely characterized through deficiencies or negative traits: a lack of fur, of natural weapons for defense; or a lack of astute sensory organs or sure instincts. To sum up, man comes into the world immature, little developed, and poorly adapted to the natural conditions of life where among the animals, those who know how to flee on the one hand and the fierce aggressive ones on the other, the human species would have become extinguished a long time ago. Nevertheless, this has not happened. Man is the master of nature and one cannot talk about the extinction of the human race. Rather, what is feared now is his excessive proliferation. How is this possible? The question and the corresponding answer have been formulated throughout the history of anthropology since Protagoras, Diogenes of Apolonia and Anaxagoras, by Pico della Mirandola, Kant, and Herder, and by contemporary anthropologists like Gehlen, Plessner, and Portmann. Through his intelligence or faculty of rational thought, man compensates for his primitive biological flaws, creating an artificial world or culture.

Contemporary philosophical anthropology, a discipline pursued primarily in German since Max Scheler [29], is based on modern biology and studies thoroughly the peculiar organic formation of man, trying to overcome the hiatus still persistent in Scheler, between *life* and *mind*, or body and soul, in terms of classical metaphysics. The cohesion between the biological and the spiritual, the emerging of the mind from nature, postulates a rationality which is not extrinsic to our animal nature as if our physical half could be reduced to this according to the traditional model. The key to the modern interpretation of man lies in the so-called "anthropina gap" ([14], p. 45) or the biological "imperfection" from which are derived the properly human characteristics. Gehlen [3] has especially insisted on this aspect. Following Nietzsche, he sees in man the "undetermined animal" because his essence has not been verified and also because he constitutes something unfinished (*unfertig*). This is a being who is lacking all the animal conditions of life and who, nevertheless, can maintain his existence. He is forced to "lead his own life" (he is, according to Herder, the "first liberated being in the Creation"; [3], p. 66) because of the position of risk and danger in which nature has placed him, depriving him of the biological guarantees possessed by the other species ("Nature is the mother of the animal and the stepmother of man" said Schelling, [3], p. 65). Because he is "incomplete", man has to realize himself: he is a being of "improvement", of permanent education. Gehlen does not hesitate to define man as *Mangelwesen*, characterized by his imperfections or defects, also called "inadequacies", "lack of specialisations", or "primitivisms" [3]. The "opening to the world", a trait so often revealed by modern anthropology, is according to Gehlen based on this defective character which makes man incapable of living in a specifically "outlined" environment. Culture compensates for man's organic primitiveness. Or it is at least founded on the human "fabric", the unusual biological "design" which, seen negatively, constitutes an incompleteness or inadequacy. Considered in a positive light, however, it signifies "creativity" ([14], p. 47), foundation for the other "anthropina" (work, progress, historicity, freedom, individuality, morality, etc.).

The concept of "biological imperfection" is closely linked to that of "sickness". Up to a certain point the *Mangelwesen*, human nature, man, as a metaphysical entity or as a biological species, must be considered to be as a "sick animal". The "pathological" element would thus define essentially the biological condition of man, exemplified if one wished in the erect posture through which the "featherless biped" demonstrates his characteristics, organic as well as moral. Here belongs the old argument of Anaxagoras and Aristotle concerning the function of the hand and the old polemic of Pindar and Galen about the centaur, renewed by Kant and Herder, in pondering the equilibrium of the quadruped as compared to the "imbalance" of the biped and in enumerating a number of ills or needs peculiarly human:

The blood that must accomplish its circulation in an upright machine, the heart crammed into an oblique position, the intestines that work in a standing container – certainly these parts are more exposed to possibilities of disturbance than in an animal body ([32], p. 335).

Leaving aside these reflections, more of an esthetic nature, Arthur Jores [8], following Gehlen, has attempted to apply the concept of *Mangelwesen* to medical anthropology, coining the phrase of "specifically human sickness." Thus one can see in the concept of "biological imperfection" a condition that causes human sickness, the "emfermabilidad", as a limitation, negation, finitude, decrepitude, the privative or defective meontological aspect that the sense of *physis* or "reality" of the pathological has primarily for man, and because of this, paradoxically, it becomes an important constitutive of human nature.

B. *Sickness and Consciousness of Alterity*

The Bible gives us the idea that self-awareness has its origin in man's ontological "flaw" (*Homo peccator*). The loss of primitive innocence coincides with the discovery of our body and our nudity (and that's why it has been said that all the anthropology is found in dressing). Conscience is a *consciousness* of limitations, originally of biological imperfections. The awakening of conscience – as it has been understood by a certain "volitional realism", represented among others by Schopenhauer, Maine de Biran, Bouterwerk, Dilthey and Scheler – is made possible through the collision with an opposition, and this implies a limitation, an imperfection. The belief that reality is felt primarily as resistance could be connected with another more general belief which sees in reflection itself – the breach between the subject and the object – an ontological negativity, a "hole in the being" [28] and even a "sickness" [33].

The phenomenological philosophy of the body recognizes the distinction between *Körper* and *Leib*, i.e., the "outer body" and the "inner body" of Ortega [20], the body-object (corporeity) and body-subject (corporality). This underlines the ambiguity between man *being* on the one hand and *having his* body on the other.

The conflict with the body, man's ambiguity in his physical existence – inasmuch as man *is* this existence and *possesses* it – appears with dialectic clarity and dramatic intensity in the experience of sickness. Man lives from his body – "Wir leben indem wir leiben" ([6], p. 118) says Heidegger – and through his body he structures all his relationships. The healthy man lives without knowing his body, hardly even conscious of it – " the simple consciousness of living" ([9], p. 117) as Kant puts it – and only in proportion to how this silent harmony, health, is broken do we become conscious of having a body that at the same time has us. The "pathological" character of the body itself has often been observed, for it "reveals itself when it rebels" with *intentio obliqua*, pointing out the dimension derived from reality as resistance, as consciousness of alterity which constitutes us as embodied existence; this manifests itself in sickness as a phenomenon of the "between" or "third nature" [7] which we are authentically, i.e., neither a pure subject nor a pure object. Indeed, the "phenomenon" of sickness, the fundamental elements of the pathological which Lain Entralgo has studied in detail [11] (*invalidity, injury, threat, loneliness, abnormality,* and *resource*), can be reduced to three origins, namely, *affection, impotence,* and *abnormality* (or the three semantic roots of the universal names of sickness: *pathos, infirmitas, nosos* or *morbus*). They point to "somatization" of conscience and the "incarnation" of existence. The whole psychological condition of human sickness can be described by means of a phenomenological analysis of embodiment [16].

C. *Sickness and the Conflict of Limitations*

The unusual human confluence of "imperfection" and "consciousness" creates a "transcendance" that we call culture and history, that never-ending process of transformation into a "second" (or "third") nature. Even when it would originally respond to the necessity of substituting for biological deficiencies, we never know if in the end transcendence aggravates these deficiencies and really signifies an incurable sickness. Indeed, civilization constitutes an "episome" or "parasome", "supplement of the soul" with which we satisfy certain necessities; at the same time it functions as an "extraneous body", promoting new necessities which in the long run increase the original dissatisfaction. It is not surprising, therefore, that human fortune is so often seen as a misfortune and *homo sapiens* as *homo demens*, man marked tragically by *hubris* or the excess of utopia. The human being, according to a pretty image of the Moslems, is like a silkworm who spins his cocoon in order to perish within his own web. Maybe man is a "sick animal" because culture has devitalized him, a "naked ape" because the culturalization of the body atrophies his original natural capacities (the use of clothing, for example, contributes to the loss of his body hair).

The philosophical anthropology which today is developing more in terms of the philosophy of history and culture ([26], [13]) has insisted on *polarity* and *conflictivity* as a permanent structure in all cultural areas, which points out a *fragility* of the systems or human "life styles", compared to the innate "ways of life" of the animal. Conflict is a specific type of *relationship*, the relationship of disharmony or discordance, precisely the Greek classical concept of sickness. A basic conflict or paradox of culture is that it constitutes at the same time the remedy and the poison for our "natural" condition. That is why philosophical thought about culture – from the Sophists and the ancient Cynics to modern "Life Philosophy" and the Utopians of today ([24], [18]) – contains attacks on culture in the name of life, and, in general, it is a historical fact that all cultures generate a "counter-culture", a rebellion of the human animal against his domestication, an insurgence of the instincts against the repressive "reason", a conflict symbolized on the relief of the Parthenon by the Apollonian gesture restraining the centaurs who, excited and drunk, burst into the society of men and threaten the established order [22].

Is is quite certain that the "diagnosis" of culture as "sickness" responds to the unilateral pondering of its negative aspects; no less certain is its "pathogenesis" or sickly character in proportion to how man cannot control and keep in equilibrium the "opening to the world", and this, instead of constituting a prosthesis for primitive biological needs, becomes the principle for the destruction of life and a threat to the human way of life. An example for this situation to which we are today especially sensitive is the ecological crisis of the last decade when nature appears to be literally "violated", devasted by technology. As Marcuse says, "Only in certain categories of sublimated aggressiveness (as in the surgeon's practice) does this violation improve the life quality of the object" ([18], p. 35). And another useful example, of psychoanalytical if not ideological inspiration, is the neurotic and alienated character of modern man who does not live in peace and harmony with his body because a harmful education (like a conjurer's to his partner in a box with a double bottom) has broken him into two irreconcilable halves, giving reality to the myth of the centaur. In whatever way, what seems more obvious today than ever before is the socio-cultural condition of that psychosomatic habit that we know as "health" and therefore the configurative role of civilization in sickness (psychosomatic influences on the habits of life – diet, work, clothing, etc. – pathogenetic factors as well as actual "models" of morbidity, e.g., arteriosclerosis and cancer). And finally, culture and history represent the "continents" of the *meaning* or meanings which men have usually given and continue to give to sickness. Lain Entralgo distinguishes between four typical and fundamental ones: punishment, misfortune, challenge, and trial ([11]. p. 186). All of these are really responses to the "mystery" of the *mal* (malaise, sickness, illness, evil) from the concrete experience of sickness, possible attitudes toward it as well as the rational *consolation*, the *contention* of the absurd or the *flight* into illusion.

III. AN ANTROPOLOGY PATHOLOGICAL

"Biological imperfection", "consciousness of alterity", and "conflict of limitations" form the "continents" or dimensions of human sickness that give it its biological, psychological and spiritual specificity or in philosophical terms its "reality", "phenomenon" and "mystery" (the human way of becoming sick is *precisely* through "nature", through "experience", and through "creativity" – qualities of *complexity*, *subjectivity*, and *plasticity* of human sickness). But those categories are not in themselves "sickness". Man's "biological imperfection" or organic lack of specialization is a negative concept that originated in comparison with the animal. But if we look at it intrinsically, the lack of specialization constitutes something positive, the basis of his very being of which deficiency is an aspect that is only privative in analogy.

Humanity, as Landmann confirms, does not represent "a way out of death-end street, a defective species that compensates for its deficiencies by creativity" ([14], p. 47). In no way does the physical worthlessness necessarily come first and the spiritual compensation afterwards, as Portmann maintains [23]. There is not, *a priori*, a *topos* of deficiency which according to Gehlen [3], following traditional thinking, is within the human body (the same shortcomings can be seen in themselves as somatic faculties). With regard to the *consciousness of alterity* (which considers resistance as a primary dimension of reality in virtue of the experience of our limitations), one could challenge it with the famous example of Kant's dove in whose flight the "consistency" of the air is as fundamental or more so than its resistance. And finally, neither does the *conflict of limitations* signify that one must consider pathological the "artificial nature of man". There is no explanation of the morbid phenomena which is not physiological or psychological, somatological *lato sensu*. (The affirmation that society acts "directly" on the body and the mind of human beings is itself unfounded and of ideological nature – which does not pretend to deny, of course, the pathogenetic and pathoplastic configuration of the historical and socio-cultural world.)

Consequently, the level of established "pathological anthropina" corresponds to what is usually called the human condition. As Zuribi puts it, "The meaning that man creates is, however, that of reality. And because it is of reality, relative to man, it reverts to human reality enriching it beyond reality" ([34]. p. 105). From this it can be deduced that "enfermabilidad", our condition of *humana infirmitas*, implies an ambivalence concerning the meaning (positive and negative at the same time) of the concrete sickness of man. Over against the "classical" concept of the pathological which has not been known to observe more than its deprived aspect, there is a "romantic" style which accentuates the creative and specifically humane side of sickness. If sickness means "imperfection", "consciousness", and "conflict" – the condition that is ours – then being man is synonymous with *being sick*. In the whole history of thought there is maybe no more eminent example of the axiological contradictions of health and sickness – properly human categories – than that of Nietzsche, himself a very sick man, which doubtlessly constitutes the only authentic form of "pathosophia".

"Of all human matters there are two" – said the physician philosopher Alcmeon de Crotona ([DK], p. 24A): health is a viewpoint on sickness and vice versa. The fundamental medical categories actually represent "facts" as much as "values", realities and axiological judgements, the latter also implicit in the epistemological structure of medicine. After all, medicine would not have a reason for existing if health were not considered to be *good* and sickness to be *bad*. Alongside the general problem that this poses – scientific knowledge and axiological prescience – medical science pursues an "objective evaluation." This rests on the axiological priority of the negative over the positive, of sickness over health, on a theory of the unreality of the positive values which derives from the basically negative character of the axiological, for those could be affirmed only through the negation of the negative values. "Disease makes health pleasant and good, hunger satiety, weariness rest" – according to Heraclitus (Fr. 111, DK). Following Maliandi [17] who has defended this theory brilliantly and has extracted all its anthropological-medical consequences, the axiological negation is not a mere subjective attitude, but rather a basic form of the experience of reality as resistance and limitation. The sick person "denies" the sickness when he values health (or the experience of sickness makes the experience of health possible). But the sickness also "denies" the sick person: it is real crisis which leads to suppressing or excluding him who suffers from it. Health is "real", but one can only experience it in contrast with sickness. Sickness, after all, is also real or rather its ontological significance is positive, although it constitutes the axiologically negative. Health for its part is positive as much from an axiological as an ontological viewpoint: only its experience of well-being and health is given in contrast with the experience of ill-being and sickness. From this the true value of sickness in the world and for man can be deduced: the intention of the "theodicies" is precisely a justification of *mal* itself, perceiving that it becomes meaningful only with experiencing well-being – "paradise is paradise lost".

The fundamental categories of medical thought – health and sickness – are therefore both ontological and axiological. But there is a question pointed out quite clearly by Gracia Guillén [5] – that of the priority of the real over the normative, of being over what should be and, in general, of "nature" over the human "condition". "The problem of embodiment and the pathological anthropology" aims at transferring the concepts health and sickness from a technical-medical level to an anthropophilosophical level, or at least to a theory of man which implies necessarily a "diagnosis" and a "treatment". The concept of man as a sick being seems to show a certain malaise in modern philosophical anthropology which can be traced to the classical dualism, which maintains its axiological root and with it a tenuous and veiled Manichaeism – that is to say, the ontologizing of sickness and the body, the incarnation of the *mal.* Without a doubt, man is a "thinking reed", according to Pascal's metaphor ([21], p. 87), "a gust of wind, a drop of water are enough to kill him". And this *horror vacui* of the physical existence is substantial with him. It is not possible to conceive the human element without this truth. But it is a "half-truth" if one persists in seeing the body only *"sub specie pathologica"*, as if it were impossible to imagine a genesis of the spirit that were not accompanied by morbid symptoms, that did not coincide with the sickness of life. In an historical sense the anthropopathological concept goes against the romantic origins of modern philosophical anthropology. From spiritualism to naturalism, sickness replaces sin in the consideration of man and his body. Without pretensions to a "metaphysical cure" – maybe impossible because of the very historicity of the *anthropos* and anthropology (Nietzsche's "perfect health" is only an ideal) – the physician-philosopher or the philosopher-physician can and must think in a healthy manner. "The healthy thought is the greatest plentitude" (Fr. 112, DK) wrote the gloomy and mournful philosopher of Ephesus.

IV. CONCLUSION

Que voulez-vous! Cela se fait un peu malgré moi. Des que je suis en présence de quel qu'un, je ne puis pas empêcher qu'un diagnostic s'ébauche en moi.... même si c'est parfaitement inutile, et hors de propos. (*Confidentiel*). A ce point que, depuis quelque temps, j'evite de me regarde dans la glace (*Knock*, Act 3, Scene IX) [25].

Knock's thesis, "Sur les prétendus états de Santé", is summarized in an epigraph attributed not by chance to Claude Bernard, "Les gens bien portant sont des malades qui s'ignorent," and finds plenty of corroboration in the theory, the technique, and the practice of actual medicine. In scientific-positive medicine, beginning with the physiopathological mentality – with Claude Bernard at the head, as has been pointed out by Canguilhem [2] – the ontological distinctions between the stages of health and sickness are erased. Against pathologic ontologism, the continuity of normal and pathological phenomena is sought within a physiological system of quantitative and gradual variations (i.e., homeostasis). Clinical judgment tends to operate under a concept of health at once *negative* (absence of sickness, "not being sick") and *relative* (referring to conventional purpose, benefit, or merit), while the concept of illness remains *positively* and categorically for nosology and nosognostics. The diagnosis of health – or the knowledge of the state of health of a person – is consequently established *per exclusionem* (via negative result of the clinical exploration; note in passing the curious semantic inversion of the terms "positive" and "negative" in clinical use such that a negative finding is a finding of health and a positive finding is a finding of disease), and *per intentionem* (via the approximation of the ideal state of physical, mental, social well-being, according to the absolute and disproprotionate concept sought by WHO). "Normal" health as normative, perfect, or ideal does not exist in practice for we are all more or less sick, as Dr. Knock suggests. The doctor will always "find something" in his patient in favor of morbid indicators. Humanity today is divided into two classes: that of the sick and the rest on whom no studies have yet been conducted: "A normal person is one who has not been sufficiently investigated" [19]. And, on the other hand, health as well-being is never fully satisfactory and is in fact a rare state.

The first and principal ethic of diagnosis becomes, therefore, conscience formed by limitations: *diagnosis* is not a *gnosis*, an intuitive and contemplative knowledge of reality, but rather essentially an operative and pragmatic knowledge in its intention (treatment), obtention (clinical exploration), and formulation (communication to the patient or third party). In addition, clinical judgment always has an evaluative aspect (in reference to values and presupposed normatives) and other *decisional* aspects (in that it sanctions a situation of social relevance, such as the *status* of sick, subject to categorical rights and obligations) as well as *interventionist* aspects (in that it responds to a therapeutic command). Therefore, diagnosis, now in its most simple concept as naming illness, implies the introduction of the subject in medical discourse, the "submission" of the patient to an order of knowledge and power. Curative or preventive, for its conservation or promotion, human life is subjected to the command of medicine: the right to health suppresses the right to illness.

As a result, to look at an individual through the eyes of medicine is to transport that individual to a terrain determined by a set of medical priorities, medical appreciations of suffering, medical appreciations of limitations. However, medical views of priorities, suffering, and limitation may not be shared with patients. Patients may have their own views of what counts as suffering, or tolerable suffering, or suffering that should be set aside by the interventions of medicine. The tension between the two perspectives becomes acute when medical classifications do not fully map onto generally experienced senses of suffering. Moreover, suffering can have personal and religious meaning. It can give significance to an individual's life and content to a culture's understanding of the human condition. The original controversies over the introduction of anesthesia for childbirth are an example. Though medicine saw suffering as something to be set aside, the culture had its reasons for affirming the suffering of childbirth.

The distinction, ontico-ontological, between illness as a nosological category and infirmitas as an anthropological category possesses a fundamental importance for the ethics of diagnosis. The confusion of the one with the other is the trap to which Dr. Knock leads us with a view of medicine as industry and commerce, instead of as an art and a science. In contrast, the infirmitas of anthropopathology permits a better understanding of "the illness of a normal man" and defends a "romantic idea" of health that includes the capacity to "be eccentric." It takes into account the sameness of human existence while respecting the rights of persons faced with the interests and purposes of medicine. An anecdote by the contemporary Austrian artist A. Kubin will serve us as a final point and a moral:

During Alfred Kubin's last illness, his doctor, about to change a dressing, turned towards his patient with a large pair of scissors in his hand. 'Have no fear', he said reassuringly. 'Nur keine Angst.' Kubin, who had spent a lifetime wrestling with the uncanny, replied: 'Don't you take away my *Angst*, Doctor. It's the only capital I have! ([30], p. 5).

La Plata National University
La Plata, Argentina

BIBLIOGRAPHY

1. Calderón de la Barca, P.: 1953, *La Vida es Sueño*, Kapeluz, Buenos Aires.
2. Canguilhem, G.: 1966, *Le Normal et le Pathologique*, P.U.F., Paris.
3. Gehlen, A.: 1958, *Der Mensch, seine Natur und seine Stellung in der Welt*, Atenaum, Bonn.
4. Gracia Guillén, D.: 1974, 'La Estructura de la Antropología Médica', in *Realitas*, I, Sociedad de Estudios y Publicaciones, Madrid, pp. 293–298.
5. Gracia Guillén, D.: 1975, 'Primer Coloquio de Humanidades Médicas', *Quirón* 3, 78 – 83.
6. Heidegger, M.: 1961, *Nietzsche*, Vol. **1**, Pfullingen.
7. Jaspers, K.: 1963, *Neitzsche*, E. Estiu (trans.), Sudamericana, Buenos Aires.
8. Jores, A.: 1960, *Der Mensch und seine Krankheit*, Ernst Klett, Stuttgart.
9. Kant, E.: 1955, *Le Conflict des Facultés* (1798), J. Gibelin (trans.) Vrin, Paris.
10. Klages, L.: 1929, *Der Geist als Widersacher der Seele*, 3 vols. Leipzig.
11. Laín Entralgo, P.: 1964, *La Relación Médico-Enfermo*, Revista de Occidente, Madrid.
12. Laín Entralgo, P.: 1968, *El Estado de Enfermedad*, Moneda y Crédito, Madrid.

13. Landman, M.: 1961, *Der Mensch als Schöpfer und Geschöpf der Kultur*, E. Reinhardt, München-Basel.
14. Landmann, M.: 1978, 'The system of Anthropina', S. Spicker (ed.), *Organism, Medicine and Metaphysics*, D. Reidel, Dordrecht, Holland, pp. 13–27.
15. Lessing, Th.: 1916, *Europa und Asien. Der Untergang der Erde am Geist*, Leipzig.
16. Mainetti, J.: 1964, 'Antropología Médica en *La Montaña Mágica* de Thomas Mann', *Quirón* 1, 55–71.
17. Maliandi, R.: 1975, 'Medicina, Axiología y Conflictividad', *Quirón* 3, 78–81.
18. Marcuse, H.: 1967, *Der eindimensionale Mensch. Studien zur Ideologie der fortgeschrittenen Industriegesellschaft*, Neuwied, Berlin.
19. Murphy, E.: 1976, *The Logic of Medicine*, John Hopkins, Baltimore and London.
20. Ortega y Gasset, J.: 1929, 'Vitalidad, Alma, Espíritu', in *O.C.*, **II** Rev. de Occidente, Madrid.
21. Pascal, B.: 1940, *Pensamientos*, X. Zubiri, (trans.) Espasa Calpe, Buenos Aires.
22. Peset, J.L.: 1976, 'Y los Centauros abandonaron la Tierra', *Quiron* 3, 16–29.
23. Portmann, A.: 1944, *Biologische Fragmente zu einer Lehre vom Menschen*, B.Schawabe, Basel.
24. Reich, W.: 1968, *La Révolution Sexuelle*, Plon, Paris.
25. Romains, J.: 1924, *Knock ou le Triomphe de la Médecine*, Gallimard, Paris.
26. Rothacker, E.: 1948, *Probleme der Kulturanthropologie*, Bouvier, Bonn.
27. Rousseau, J.J.: 1977, *Discursos a la Academia de Dijon*, A. Pintor Ramos, (trans.) Ediciones Paulinas, Madrid.
28. Sartre, J.P.: 1943, *L'Être et le Néant*, Gallimard, Paris.
29. Scheler, M.: 1928, *Die Stellung des Menschen in Kosmos*, Otto Reichl, Darmstadt.
30. Sebba, G.: 1973, *Kubin's Dance of Death*, Dover, New York.
31. Spicker, S.: 1976, 'Terra Firms and Infirma Species: from Medical Philosophical Anthropology to Philosophy of Medicine', *The Journal of Medicine and Philosophy* **2**, 104–135.
32. Straus, E.W.: 1970, 'Born to See, Bound to Behold: Reflections on the function of Upright Posture in the Esthetic Attitude', in S.F. Spicker (ed.) *The Philosophy of the Body*, Quadrangle, New York, pp. 334–361.
33. Unamuno, M. de: 1950, *Del Sentimiento Trágico de la Vida*, Espasa Calpe, Buenos Aires.
34. Zubiri, X.: 1962, *Sobre la Esencia*, Sociedad de Estudios y Publicaciones, Madrid.

DREW LEDER

THE EXPERIENCE OF PAIN AND ITS
CLINICAL IMPLICATIONS

Diagnosis is an ontological event. That is, to arrive at a diagnosis is not simply to solve a puzzle; it is to ratify a vision of reality. When the words "peptic ulcer disease", for example, are pronounced, the ambiguous complaints of the patient – gnawing pain, hunger, and the like – are reorganized into a medically defined world. Events take on a new spatial contour: they now center around and radiate from the duodenum. Causality is given a mechanistic reading. It is physiological mechanisms – the secretion of acid, the erosion of tissues – that order this world. As diagnosis goes hand-in-hand with prognosis, a projected future is brought into being. Moreover, a certain course of action is prescribed, as diagnosis helps determine the proper mode of treatment. The symptoms of the patient are thus captured within a structured world-view, with spatial, temporal, causal, and prescriptive dimensions.

The fundamental outlines of this world-view are held constant throughout the range of diagnoses available to the modern physician. For all share a basic commitment to the physiological reading of disease. Whether the diagnosis is peptic ulcer or a carcinoma of the stomach, causality is understood in mechanistic terms. There is no explicit reference within this world to evil demons or sin and punishment. The primary locus wherein sickness unfolds is seen as the isolated body of the patient, not a broad social or cosmic matrix. And this body is not a magical or experiential field: it is a complex machine, specified according to anatomical and mathematical parameters. When a diagnosis is made, even one as minor as "acne vulgaris", the entirety of this world-view is implicitly present.

But this world of physiological diagnosis is not identical to the world of the suffering patient. While the doctor attempts to understand and treat a *disease*, the patient is undergoing an *illness* ([5], [8], [15], pp. 441–445). An "illness", according to this terminology, involves the symptoms and life-transformations experienced by the sick. In the case of the peptic ulcer patient, this might include pain and a consequent surly disposition, a loss of sleep, an inability to eat desired foods, fear and confusion regarding what is taking place, an impatience with others who

95

J. L. Peset and D. Gracia (eds.), The Ethics of Diagnosis, 95–105.
© 1992 *by Kluwer Academic Publishers. Printed in the Netherlands.*

seem oblivious. The patient's sense of time and space, of embodiment and sociality, have all been disrupted and reorganized. It is as a result of such experiences that help is sought.

Hence, the clinical encounter involves a meeting of two worlds, one defined in terms of physiological diseases, the other through illness experiences. There are, indeed, world-bridging structures: the compassion of the physician and the scientific awareness of the patient both facilitate communication back and forth [8]. The success of the therapeutic encounter rests on achieving a certain degree of shared understanding. However, these two worlds never fully overlap. The physiological world-view in which the modern physician is trained involves a language, style of thought, and set of categories, which differ markedly from those of the suffering patient.

It is not merely that the physiological world-view omits certain experiential concerns; more pointedly, it can serve directly to obscure them. Prior to the nineteenth century, diagnostic categories were based largely upon the reported symptoms of the patient. However, as Foucault ([10], pp. 124 –148), Reiser [14], and Engelhardt ([9], pp. 176–184) have discussed, with the rise of the pathoanatomical method, such symptoms came to be seen as somewhat epiphenomenal; diseases were re-classified according to lesions found in the anatomized corpse, or in the living patient via diagnostic technologies. The patient's experience becomes increasingly peripheral when "truth" is to be uncovered in the pathologist's laboratory, or through X-ray and blood chemistries [2].

On the one hand, this shift from a focus on symptomatology to a focus on underlying disease mechanisms has had enormously beneficial results. Much of the efficacy of modern medicine can be traced to this conceptual reorganization. On the other hand, much has been lost as well. Many of the widely-recognized failures of modern medicine – depersonalized treatment, overreliance on medical technologies, non-compliance on the part of patients – relate to this eradiction of the patient's voice, this neglect of illness in favor of disease.

In this essay I will address the illness experience, via one of its most common and devastating features: the phenomenon of *pain*. First, I will examine the way in which pain transforms one's relationship to the body, and to the surrounding world. I will not attempt anything like a comprehensive phenomenology of the pain experience. I have taken up this topic at greater length elsewhere [12], and there are several excellent studies in the literature ([1], [3], [16]). I will only briefly trace out one dimension of pain: its tendency to establish opposition and duality. Secondly, I will discuss the clinical implications of this analysis. If pain engenders experiential *duality*, a crucial function of the medical encounter must be to restore the lived *unity* of the patient. Seen within this context, it is not only technical treatments that have therapeutic efficacy; the doctor-patient relationship, even the process of diagnosis itself, manifests a power to relieve suffering.

I. THE EXPERIENCE OF PAIN

A primary theme, almost an obsession, of modern philosophy, has been the attempt to define the mind-body relationship. Descartes, most famously, proposed a dualistic solution. Mind and body, though conjoined within the person, are understood as separate and essentially distinct substances. Such an account is profoundly challenged by the experience of pain. For in pain we find the unity of the mental and corporeal. The bodily sensation is intimately intertwined with an emotional and existential meaning. This unity has a linguistic representation: the word "pain" itself can be used to describe not only physical, but emotional or cognitive suffering. The relationships expressed by this multivocal word are more than analogical. The pain of grief, for example, is not "merely" emotional, but bound up with a gut-wrenching physical expression. Conversely, physical pain is rarely devoid of emotion: it bears within it a component of displeasure and aversion, and often of anxiety, sadness or rage. Cognitive significance is equally immanent within the physical: the pain experienced by a patient with a mysterious tumor is alleviated when a benign diagnosis is pronounced and intensified by that feared word, "cancer". This is not only a change in intellectual interpretation. There is a profound transformation in the experienced pain itself.

Pain not only reveals the unity of body and mind, but of self and world. Philosophers have often regarded pain as the quintessential private experience, localized within the confines of one's own consciousness. However, pain is never simply a sense-datum, an interior thing: it reverberates throughout the experiential world. When in pain, those activities that yesterday held enchantment, today are revealed as irrelevant or repugnant. Spatiality is no longer expansive. It now constricts around our bedroom and the affected body part. As we are drawn back to the "here" of the suffering body, so also to the "now": pain rivets us to a present from which we wish to escape. The person in pain thus undergoes an alteration of world. Suffering is not measured in C-fiber firings, but in the contraction of reality and possibility.

Hence pain never simply adheres to body rather than mind, self rather than world. Dualistic models tend to obscure the essence of this phenomenon. Yet, at the same time, pain does engender a profound experience of duality. Pain fractures our habitual universe, creating a series of lived oppositions. That to which we are most intimately connected becomes that from which we are most estranged.

I will discuss this oppositional force of pain first in relation to mind and body. As even Descartes acknowledged, pain reveals one's intimate union with the body in a powerful way ([7], p. 192). Yet it does so via a mode of alienation. In our daily life we simply *are* our body, moving through the world as an integrated whole. But when beset by pain, the body surfaces as Other, something disharmonious with the mind ([17], p. 245). The body that once was silent now screams for attention. Where once it obeyed the least command, it now rebels against our desires and undermines our projects. We would be rid of this suffering, but find ourselves subject to a foreign will. At such moments our corporeality presents itself as both vulnerable and a threatening thing, a place of imprisonment. The modern medical encounter is often criticized for alienating and objectifying the body. However, this only extends a process begun by the illness [11]. For the painful body is profoundly alien. Before coming to the doctor, the patient has already begun to objectify her body, poke and prod it, wonder at and resent its strange autonomy. The mind thus experiences the Otherness of the physical.

Just as pain can alienate mind from body, so it can separate self from its surrounding world. This is, once again, a lived paradox where connection and opposition are simultaneously asserted. Pain undermines any fantasies of pure transcendence, of a self detached from its social and material conditions. Pain recalls us to our finitude and dependency, dragging us back into the mundane world. Yet this is a world in which we no longer feel at home. This physical or social matrix has brought about suffering. Hence, the person in pain may choose to retreat into the privacy of the bedroom or the oblivion of sleep.

Moreover, others frequently retreat from the sufferer. The pain of another is a source of empathic suffering and fear, and an acute reminder of our vulnerability. Even if family and friends remain solicitous, the person in pain discovers the limited nature of the help they can provide. Others can neither experience one's pain nor remove it. The sufferer often seeks companionship in a heightened way, but this relates to a heightened sense of distance. In the words of John Updike:

(Pain) shows us, too, how those around us
do not, and cannot, share
our being; though men talk animatedly
and challenge silence with laughter

and women bring their engendering smiles
and eyes of famous mercy,
these kind things slide away
like rain beating on a filthy window

when pain interposes [18].

Pain forces itself between self and other, opening an existential rift. This is intensified when chronic pain renders us unable to participate in customary social roles. We no longer go to work or join in the recreation of others. The everyday world continues – we hear it poignantly from our bedroom window – but it goes on without us ([19], pp. 23–27). The isolated subject is left, a proto-solipsist. Although pain is not metaphysically private, it brings with it privation. It establishes an experience of isolation, of the distance separating self from world.

II. CLINICAL IMPLICATIONS

Physical pain is ingredient in many, though not all, of the situations which bring a patient to treatment. Moreover, most sicknesses, even if not dominated by pain, exhibit a similar phenomenological structure. For example, a chronic heart condition may progress without a great deal of pain. Yet there is still a sense of exile and isolation: the individual is singled out from others by this life-endangering condition. Nor is the body any less transformed into alien object. Even when producing no immediate discomfort, the diseased heart remains a threat to the self's very existence. Illness thus dissolves the harmonies upon which our everyday life is built. This is the result, depending upon the condition, of discomfort, disability and/or disfigurement.

To treat an illness, not simply a disease, is to participate in the restoration of harmony. The healing task of the physician, or any health care provider, thus cannot be understood purely in physicalistic terms.[1] This is not to deny the crucial importance of the doctor's technical armamentar-

ium. A pill or surgical procedure that abolishes the source of pain will do
more to restore the life-world of the patient than any amount of sympa-
thetic concern. As a disease is healed, the experienced rift between mind
and body, self and world, heals as well. To oppose the technical to the
humanistic aims of medicine is to falsify their unity in the therapeutic
encounter.

However, the technical resolution of pain and sickness is hardly the
only mode of therapy operative in medicine. With most of the chronic
diseases that plague contemporary society – cancer, heart disease, arthri-
tis, bronchitis-emphysema – there are no simple curative procedures.
Pain and disability are an ongoing fact of life to be managed and amelio-
rated, not removed. In other situations where pain is more easily abol-
ished – for example, a pneumococcal pneumonia responsive to penicillin
– there is always some interim period before the treatment regimen takes
effect. Yet the healing process can and should begin immediately. For
inherent in the doctor-patient relation is the potential to alter the phenomenology of suffering. Here can begin the reconstruction of all that
pain has deconstructed.

We have seen, for example, the division pain introduces between self
and others. This division exhibits what I will term an *epistemic* and a
praxical dimension. Pain poses an epistemic block by virtue of its indi-
vidualizing function. That is, another is unable to fully know or understand the experience of the sufferer. Moreover, along the praxical
dimension, the ability of others to help is curtailed. In facing external
threats, I can call upon friends and family for concrete assistance. Many
of life's difficulties pose a communal challenge and yield only to a communal resolution. But pain must largely be borne alone. The sufferer discovers the limits of assistance.

The act of seeking medical care can play an important role in reversing
such trends. Along both epistemic and praxical lines, the health care practitioner is in a unique position to maximize the possibilities of intersubjectivity.

I will first discuss this in relation to the *epistemic* dimension: the
accomplishment of knowledge and understanding. Mutual understanding,
in a vital sense, begins with the expression of compassionate concern for
the patient. The word "compassion" means, etymologically, to "suffer
with". To thus suffer with the affected person helps relieve pain's isolation. Family and friends may have already begun such a process.
However, a medical training allows the treater to extend this process in

crucial ways. By virtue of a constant exposure to the world of illness, clinicians can cultivate a deepened understanding of pain, of its capacity to ravage one's psyche and world. Moreover, they are trained to listen to the particular suffering of *this* individual. In taking a history, the story of the pain is told in a detail few others would tolerate. Beyond its diagnostic import, there are therapeutic ramifications to this narrative form. As pain reaches its maximal expression in language, the solitude it imposes is partially overcome. Whereas pain is the most interior of sensations, language externalizes and accomplishes a sharing ([16], pp. 3–23). Finally, the clinician does not simply hear the words, but can understand them in ways that others cannot. Each doctor is a *hermeneut*, interpreting the patient's language and bodily state in order to uncover a hidden meaning: the disease process that lies beneath [6]. The clinician thus comes to *know* the pain, its source and significance, and can share this knowledge with the patient. An epistemic bridge is built. In this restoration of intersubjective understanding, the relief of pain has already begun.

Moreover, this epistemic bridge between self and world is reinforced by a *praxical* dimension. The practitioner does not just understand: his or her understanding is the tool that facilitates intervention. The helplessness of the world to relieve one's pain is reversed, as much as possible, through the figure of the healer. As a result of training and a technical armamentarium, the practitioner is able to reach into the interior of another's body to effect concrete assistance. If pain cannot be removed it can usually be relieved. The medical relation reasserts the import of the other over against pain's solipsism.

The power of the medical encounter to restore sociality extends beyond the individual doctor-patient relation. The clinician can lessen disabling symptoms and provide assisting devices: this, in turn, may help restore the patient to his or her customary social context. Moreover, by educating family and friends concerning the diagnosis, therapy, and needs of the patient, the practitioner can help reconstitute the epistemic/praxical power of this social network. Others are then better able to understand and render assistance to the sufferer.

The contemporary medical care system does not always foster this restoration of world. A callous or uncommunicative doctor can heighten rather than relieve the patient's sense of isolation. Moreover, the process of hospitalization, while sometimes necessary, often exacerbates the experiential dislocations of pain. The hospitalized person is removed from friends, family, and habitual surroundings. Even one's clothes, a marker of social identity, are taken away. One enters the hospital a stranger in a strange land. If illness creates a gap between self and world, this gap may be widened by such forms of treatment. We might term this "iatrogenic illness", to be distinguished from iatrogenic disease.

However, this need not be the result of medical care. We have discussed the power inherent in the therapeutic relationship to counteract the self-world division imposed by pain. The same is true in relation to the rift between mind and body. There is, once again, an epistemic and a praxical dimension to this rift. The person in pain often suffers from a lack of knowledge. Bodily processes, heretofore taken for granted, now surface as mysterious and inexplicable. The patient seeks answers concerning the physical domain. Moreover, pain not only initiates a crisis in the sphere of knowledge, but of praxis. Ordinarily one exerts unproblematic control over one's body. But the painful body stands in opposition to the mind, resisting its projects and attempts at relief.

The practitioner, again, is in a unique position to mediate between conflicting terms. The *epistemic* grasp of diagnosis and prognosis can be communicated to the patient. Even if the message is not a pleasant one, most people will experience some relief at finally knowing. Bodily events are not then as foreign to mind. Diagnosis thus has therapeutic import, helping to relieve a cognitive crisis.

Moreover, the practitioner is capable of returning to the patient, as much as the clinical situation permits, the element of *praxical* control. Through the medium of the health care professional, the patient learns to alter or cooperate with physical changes rather than remaining their helpless victim. This is enhanced when the patient's decisional power over the treatment regimen is maximized. Mind regains some mastery over the body. The crisis of powerlessness instituted by pain is partially, if not totally, overcome [4].

III. CONCLUDING ETHICAL REFLECTIONS

Close attention to the experience of pain thus sheds new light on medical therapeutics. Treatment is more than correcting physiological abnormality: it is the process of reconstructing a shattered domain. Similarly, the field of medical ethics might be transformed by a phenomenological understanding of illness. In closing, I will briefly suggest the import of our analysis vis-á-vis contemporary ethical debates.

Many of the debates within medical ethics center around the question of patient autonomy. This includes, among others, issues of truth-telling and paternalism in the doctor-patient relationship, informed consent and the right to refuse treatment, confidentiality, genetic screening, involuntary commitment, and euthanasia. Such discussions often assume a classical ethical stance in which the person is viewed first and foremost as an abstract moral being. That is, the patient is regarded as a universalized self, a denizen of the Kingdom of Ends, with fundamental rights and freedoms.

On the one hand, such a framework assures due respect for the person weakened and compromised by illness. On the other hand, it tends to obscure the crucial role of pain and illness in shaping the ethical dilemmas of medicine. The sick patient seen in clinical practice is never simply the autonomous self of moral theory. His or her selfhood and autonomy are precisely what illness has called into question.

Our previous analysis has sought to illustrate this point. For the person in pain, selfhood is rendered experientially problematic. Mind and body are split asunder and one's physical integrity is thrown into doubt. Stripped away are the social roles upon which customary identity is based. One's sense of self is thus profoundly challenged. This is true as well of one's sense of autonomy. Autonomy requires a certain understanding of events and of the range of possible responses. But we have seen that pain brings epistemic crisis: one often knows neither the meaning nor causes of the affliction. Nor can one exert effective control. Autonomy is eroded by pain's brutal regime.

In this light, the traditional debates of medical ethics take on a new meaning. For example, the principle of "informed consent" can no longer be viewed simply as a way to safeguard a pre-existing autonomy. Rather, it may serve to rebuild an autonomy already torn apart by pain and sickness. In securing truly informed consent, one provides the patient with knowledge of and control over his/her situation, while reinforcing a sense of intersubjective trust. Knowledge, control, and intersubjective trust: these have all been undermined by pain. For this reason, informed consent is not simply a preliminary to therapy, but is itself a therapeutic principle.

Hence, not simply the abstractions of moral theory, but the realities of illness, suffering, and disability constitute the proper ground for medical ethics. ([13], pp. 170–200; [20]). To recognize this is not to call for a single overarching phenomenological-based system to replace the grand theories of normative ethics. Rather, there is a need to conduct regional studies appropriate to the specific moral dilemmas of health care. This analysis of pain represents one such study, but it is possible to envision many others. For example, the debate over involuntary psychiatric treatment needs to be situated via the phenomenology of psychiatric illness – the way such disturbances may already have impaired one's sense of volition and one's confidence in others. Similarly, debates concerning voluntary euthanasia remain abstract when removed from the experiential context. This includes the world-reduction effected by chronic illness, the contraction of temporality and expectation which accompany the diagnosis of a fatal disease.

This is not to say the phenomenological inquiry will unambiguously resolve questions of right and wrong. That can only be done by a formulaic and, hence, unsatisfactory ethics. But without attending to the lived experience of illness our moral analysis remains incomplete. For the ethical dilemmas of health care are existential dilemmas as well. They arise in a context of profound suffering and hope, threat and possibility, isolation and compassion. This need be the living reality of the bioethicist no less than that of the practicing clinician.

Loyola College in Maryland
Baltimore, Maryland

NOTE

[1] I will from time to time, for convenience, refer to the "physician" or "doctor". However, my analysis is meant to apply to health care professionals in general, insofar as they participate in the diagnostic and therapeutic process.

BIBLIOGRAPHY

1. Bakan, D.: 1971, *Disease, Pain, and Sacrifice*, Beacon Press, Boston, Mass.
2. Baron, R.: 1985, 'An Introduction to Medical Phenomenology: I Can't Hear You While I'm Listening', *Annals of Internal Medicine* 103, 606–611.
3. Buytendijk, F.J.J.: 1962, *Pain, Its Modes and Functions*, E. O'Shiel (trans.) University of Chicago Press, Chicago, Illinois.
4. Cassell, E.: 1977, 'The Function of Medicine', *Hastings Center Report* 7, 16–19.
5. Cassell, E.: 1985, *The Healer's Art*, MIT Press, Cambridge, Mass.
6. Daniel, S.: 1986, 'The Patient as Text: A Model of Clinical Hermeneutics', *Theoretical Medicine* 7, 195 210.
7. Descartes. R.: 1911, *The Philosophical Works of Descartes*, Vol. 1, Cambridge University Press, Cambridge, Mass.

8. Engelhardt, H.T., Jr.: 1982, 'Illnesses, Diseases, and Sicknesses', in V. Kestenbaum (ed.), *The Humanity of the Ill,* University of Tennessee Press, Knoxville, Tennessee, pp. 142–156.

9. Engelhardt, H.T., Jr.: 1986, *The Foundations of Bioethics,* Oxford University Press, New York.

10. Foucault, M.: 1975, *The Birth of the Clinic,* A.M. Sheridan Smith (trans.), Vintage Books, New York.

11. Leder, D.: 1984, 'Medicine and Paradigms of Embodiment', *The Journal of Medicine and Philosophy* **9**, 29–43.

12. Leder, D.: 1984–85, 'Toward a Phenomenology of Pain', *Review* of *Existential Psychology and Psychiatry* **19**, 255–266.

13. Pellegrino, E., and Thomasma, D.: 1981, *A Philosophical Basis of Medical Practice,* Oxford University Press, New York.

14. Reiser, S.: 1978, *Medicine and the Reign of Technology,* Cambridge University Press, Cambridge, Mass.

15. Sartre, J.P.: 1966, *Being and Nothingness,* H. Barnes (trans.) Washington Square Press, New York.

16. Scarry, E.: 1985, *The Body in Pain,* Oxford University Press, New York.

17. Straus, E.: 1963, *The Primary World of Senses,* J. Needleman (trans.), The Free Press of Glencoe, New York.

18. Updike, J.: 1983, 'Pain', *The New Republic* **189** (Dec. 26) 34.

19. van den Berg, J.H.: 1966, *The Psychology of the Sickbed,* Duquesne University Press, Pittsburgh, Pennsylvania.

20. Zaner, R.: 1984, 'Is "Ethicist" Anything to call a Philosopher?', *Human Studies* **7**, 71–90.

STUART F. SPICKER

ETHICS IN DIAGNOSIS: BODILY INTEGRITY,
TRUST-TELLING, AND THE GOOD PHYSICIAN*

I. THE INTERTWINING OF DIAGNOSIS AND TREATMENT

In addition to the acquisition of differential knowledge on the part of a physician, medical diagnosis depends on the subtleties of the patient-physician encounter, the prevailing nosological and conceptual systems in which this encounter occurs, the current status of biomedical research, and the physician's reliance on the accuracy, reliability, and validity of the available laboratory analyses. Indeed, one can also make the case that the process of medical diagnosis tacitly involves the normative presuppositions of the medical context, the uncertainties of questionable diagnoses that may lead to ill-effects (including the physician's failure to obtain peer and family approval rooted in a lack of confidence in his or her diagnostic competence), the values of other health care practitioners involved in a patient's case, the diagnostic modalities available in the clinical setting in which specific medical acts take place, and even *post-mortem* procedures [8] (when no treatment is possible) as propaedeutic to future diagnoses that are often forgotten but critical to public health problems like today's AIDS epidemic.

It should be clear from the outset that *screening* differs from conventional, invidual diagnosis; the former applies to populations at *risk* but without specific complaints that would lead to specific testing. Diagnosis, however, is directed to the future care of an individual patient, and is usually initiated by the acquisition of data which suggests specific disease processes ([15], pp. 25–26). However, both diagnosis and screening can be used (1) to establish cohorts of research subjects (statistical aggregates), or (2) to acquire baseline information on the idiosyncratic nature of individual patients, although such information remains confidential, and requires physicians to consider the privacy of their patients.

An examination of the ethical issues ingredient in patient care – beyond privacy and confidentially - reveals the fact that medical diagnosis, treatment (and prognosis) continuously interact, for essential aspects of treatment are often, but not always, introduced into the core of the diagnostic

107

J. L. Peset and D. Gracia (eds.), The Ethics of Diagnosis, 107–122.
© 1992 *by Kluwer Academic Publishers. Printed in the Netherlands.*

process. That is, once the physician is clear on the patient's chief complaint – the "nodal point" – treatment may be initiated during diagnosis when adequate diagnostic information may not yet be available, though the process of diagnosis frequently continues after treatment has already been initiated [12], since important information may or may not be introduced at a later time, "the truth of the diagnosis [being] ... almost a matter of opinion".[1] For example, a patient may be quite febrile, and the physician may prescribe aspirin or another agent to provide *symptomatic* relief before knowing the etiology of the disease. Again, she may also initiate a therapeutic trial to confirm or reject a diagnosis. For example, a patient may be suffering from what the physician knows broadly to be X. The physician would like to know which form of X the patient has, in order to prescribe a medically proper course of treatment. But the physician also knows that patients suffering X are typically agitated and anxious, but tend to respond well when they feel cared for, that X itself is partially relieved by such care. In this case, too, diagnosis and (placebo) treatment are intertwined. Thus it is important to eschew the "standard view" that diagnosis is typically a process separate from treatment.

The intertwining of diagnosis and treatment in the form of a "course of action" is more obviously seen by turning to the literature on medical decision-making. Among their "clinical aphorisms", Dr. Harold C. Sox, Jr., *et al.*, include the following pithy distillate: "If a test is unlikely to change the management [treatment] of the patient, don't do the test" ([19], p. 25). Is this maxim value neutral? At its most straightforward reading it may be proffered in order to signal the need to conserve resources and to avoid costly tests where possible. Or does it admit of a more normative interpretation? That is, the patient's well-being and status require that *risks* to the patient be kept to a minimum even during the diagnostic process, though it is a commonplace that such risks are frequently undergone in the treatment phase. Thus, notwithstanding the fact that the process of differential diagnosis requires the generation of alternative hypotheses, the gathering of data, and the use of the data to test the various hypotheses, the *final step* in the process of clinical reasoning to determine the causes of the patient's complaint is the transition to treatment – "select a course of action" ([19], p. 26).

It is also possible to identify a number of normative issues ingredient in *diagnosis* that should be distinguishable from those ethical issues that emerge from an analysis of medical practice, i.e., the myriad of ethical issued tacitly involved in treatment (management) or nontreatment decisions. I am thinking here of what I shall call the "internal norms" of medical diagnosis that are analogous to the internal norms of scientific inquiry: physicians expect each other (1) to work to overcome observer bias, (2) to tell the truth and be honest with those with whom they communicate, (3) to report accurately the results of their investigations and case analyses in the literature, (4) to be mindful of what the public may view as inappropriate or inordinate financial gain by periodically evaluating, for example, personal investment/ownership in local diagnostic laboratories, (5) to gather evidence accurately, that is, to search for "truth" in medical diagnosis, and (6) take all patient reports – concerning how he or she feels – seriously; for as I shall soon argue, "the reasons for listening to patients are not only moral, they arise from the nature of diagnosis itself" (a point I owe to Roger Crisp).

Traditionally, the process of diagnosis was not the place one would have expected to discover ethical problems beyond the "internal norms" noticed above, though contemporary Foucauldian literature reveals concern over the fact that the physician, through acquiring knowledge, "serves as judge" and thereby gains *power* over his patient, power that can be used to eliminate a patient from certain social categories, e.g., employment, or to discriminate against or control a patient by informing the law of, e.g., his sexually contagious or lanthanic genetic disease.

These and similar concerns typically emerge when various treatment modalities are being evaluated. When physicians outline treatment plans for many of their patients, the patients are often well enough to take such occasions to accept or reject the proposals. Indeed, today it is expected that if he is capable, the patient will evaluate the explanation his physician is required to provide, especially the proposed treatment plan (that may prove quite risky), and also to participate in the decision-making that may follow; he may even elect to postpone or suspend such decisions, however, and request (demand?) further medical or ethical consultation.

One of the few places in the extent bioethical literature on diagnosis where one can find a "cursory glimpse" of the ethics of or *values* [2] ingredient in diagnosis is K. Danner Clouser's digression in "Approaching the Logic of Diagnosis" ([17], pp. 160–176; see also [3], [12]). Here Clouser employs the term 'values' not 'ethics', since he is merely noting the *normative* ingredient in the process of differential diagnosis. He observes that since the outcome of differential diagnosis must be *acted* upon, "we have no alternative but to weigh carefully the value of what we have to win or lose against the odds of winning or losing" ([17], p. 144). Thus, in the process of deciding, for example, what laboratory tests to order, the physician is *implicitly intending a preference* toward the end of assisting the patient.

It should be obvious that this step naturally involves the weighing of *risks* (as well as benefits) – pains suffered, expense, delay of treatment, worry, and possibly contagion and loss of life for the individual patient. After all, as John Fletcher has reminded me, a number of patients, especially infants, are put at *significant risk* during the diagnostic process. In spite of various "imaging technologies", many diagnostic tests are so invasive that in the very process of acquiring information the patient may be infected, injured, or even killed. And as Dr. Henrik R. Wulff has understated it: "... some diagnostic procedures may lead to complications" ([24], pp. 5–6); but furthermore, he points out," ... it may be suspected that patients are subjected to many unnecessary investigations".... Often "we have not proved their ultimate benefit to our patients" ([24], pp. 111, 115).[3] Even more worrisome, as Robert U. Massey, M.D., has suggested, is the frequency with which today's physicians prescribe a test or a series of tests for a prospective patient *prior* to a careful history-taking, physical examination, and use of logic. The rationale frequently offered is that technological procedures and their reported results will "save time" or "save money" (perhaps thought?), since physical examination takes the physician's time and tests do not ([16], p. 1632); indeed, such practices typically increase costs when laboratory findings are substituted for clinical judgment, i.e., diagnoses, and not infrequently increase risks for the patient. Clearly there is a serious normative concern: the "un-met" patient may be placed at significant risk precisely because the physician does not know the patient's state of health or infirmity. The patient is "an unknown", as well as unknown in the ordinary sense. An extension of this pernicious practice is making "chart rounds" where one avoids seeing the patient. This is reminiscent of the 17th-century practice where physicians made a diagnosis based on a horoscope before seeing the patient!

It is on the basis of such risks that physicians are appropriately concerned with misdiagnoses and false negative claims and tests (so-called Type II errors). Such worries are "shot through" with value, i.e., what is good or bad for the patient (which, of course, should be distinguished from the actions a physician may taken merely to avert the creation of an unfavorable impression toward him or herself).

II. TRANSITION TO ETHICS IN DIAGNOSIS

At this point one may simply mention a few specifically moral issues pertinent to physicians' *decisions to treat*: (1) issues that concern the patient's responsibility following the *act* of diagnosis, when he or she is also expected to make some treatment decisions in his or her own case; (2) issues concerning the extent to which patients should be informed about medical diagnosis in general, as well as thoroughly informed of their own case during the initial stages following the diagnostic process – the issues ingredient in "informed consent" ([12] pp. 137–140); (3) issues that occur when it is argued that patients have a moral right to interrupt the process of diagnosis; and (4) issues of treatment that intertwine with and complement the diagnostic issues – "Should the physician consult the computer ([8], [17], pp. 47–52), withhold or withdraw treatment, or act on the clinical judgment that to do nothing for the moment is the best treatment of all?" In short, ethical judgments pervade all clinical decision-making – "even the simplest decisions on the daily ward round" that create a risk of negative side-effects and lower quality of life ([24], p. 160).

Notwithstanding this set of ethical issues germane to treatment, it is not uncommon to find that standard texts on medical diagnosis and decision-making omit analysis of even the most obvious ethical issues integral to differential diagnosis, although such texts usually include cursory remarks like "The goal of clinical decision-making", for example, "is to maximize the patient's well-being". To do so, the authors continue, "the clinician should try to take into account the patient's judgment about the outcomes that may be experienced". Having done so, the principal goal is to "try to measure" these experiences, a difficult task to be sure, because "[f]eelings about the experiences of illness are so subjective and personal" ([19], p. 168). Not surprisingly, another medico-scientific nod to objectivity over subjectivity!

Now, by attending to what physicians decide to do (based on framing diagnoses and engaging in responsible medical decision-making) followed by what, in fact, they carry out through various treatment plans, it soon becomes apparent that important non-medical e.g., evaluative, normative, or ethical – factors frequently influence the *treatment* prescribed for patients. For example, given a diagnosis of non-small-cell lung cancer, different patients are often given different treatments – possibly for reasons independent of their clinical condition and the physician's uncer-

tainty regarding the optimal therapy – i.e., some may be treated by surgical procedures (thoracotomy with lobectomy, or with pneumonectomy)
and some not surgically treated, based on current functional assessment,
age, comorbidity, or even socioeconomic status – a clearly ethical (non-
medical) criterion ([4], [8]). However important such factors are, they
often serve to camouflage the ethical issues of *treatment*, and they may as
well serve to mask those ethical issues that are directly due to the *process
of diagnosis* ([13], [10]). Nowithstanding the fact that treatment choices
under conditions of uncertainty [6], even where such choices have
significant implications for the life and life styles of patients and frequently reflect the presence of a normative or ethical component, physicians continue to witness an excessive attention to this domain that has
tended to make them and others neglect the ethical issues in the process of
diagnosis. This is partially understandable because, as Drew Leder
remarks, "Each doctor is a hermeneut, interpreting the patient's language
and bodily state in order to uncover a hidden meaning: the disease process
that lies beneath. The clinician thus may come to *know* the pain, its source
and significance, and can share this knowledge with the patient. An epistemic bridge is built" ([9], p. 101). Or, as other clinicians have remarked,
"Simply letting the patient recount the problem usually provides sufficient
information to narrow the field immensely in a short time" ([1], p. 201). In
the process of diagnosis itself, then, may be found the initial (normative)
elements of relief, and therefore it is generally true that "clinicians remain
the best diagnostic instrument" ([1], p. 189).

III. EMBODIMENT, ETHICS, AND THE ACT OF DIAGNOSIS

Based upon the distinction between the ethics of diagnosis *and* the ethics
of treatment, in what follows I shall remain fixed on the former. Rather
than ask, for example, what are the ethical implications of a physician's
classification of types of cancer – anaplastic, epidermoid, adenocarci-
noma; or, what a particular staging of cancer means in terms of the
physician's formulation of a treatment plan, I shall ask: How does a
patient's image of his embodiment become transformed upon an accurate
diagnosis of serious illness – what is the meaning of the loss of particular
"body parts" for the patient? Further, what specific ethical issues are
ingredient in this complex process *prior* to medical or surgical interventions? That is, (1) in what way, if any, does the process of medical diagnosis conjoined with a descriptive phenomenology of the patient's lived

body during serious illness suggest that *physicians* ought to enter into a *special trust* relation with the patient, and (2) how may this process of "trust telling" affect the *patient's* apprehension and appreciation of his future incapacities and anticipated functional losses following the diagnosis and even the initiation of treatment? After all, patients suffer a myriad of such losses: loss of *agency*, loss of *control*, loss of various *capacities* (e.g., locomotion [5], hearing, seeing, etc.).

To pursue this two-part question, therefore, *the process of diagnosis* and not (what I prefer to call) the *act* of diagnosis nor the initiation of treatment must remain salient. I shall, however, presume that the "ideal" physician has involved his or her patient in the diagnosis, *not perhaps by exploring and explaining the process of diagnosis in detail*, but by sharing with the patient the information and "name" of the disease condition, i.e., the *act* or *outcome* of diagnosis. [4] Thus a diagnostician carries out an act of diagnosis when he or she communicates the nosology to the patient, though such a precise temporal moment does not occur in all cases. Leder remarks: "Even if the message is not a pleasant one, most people will experience some relief at finally knowing.... Diagnosis thus has therapeutic import, helping to relieve epistemic crisis" ([9]. p. 102). And in her, *All of a Piece: A Life with Multiple Sclerosis*, Barbara D. Webster relates the following:

My first reaction to receiving a diagnosis of multiple sclerosis was one of shock and terror, mixed with a deep sense of relief. I was stunned and I knew very little about the disease, which added to my terror, but at the same time there was overwhelming relief in knowing there was a solid physiological reason for the symptoms and inexplicable bouts of illness I had experienced over the [past 14!] years ([23], p. 23).

In short, the *actual* moment of communicating the name of the disease by way of clear and comprehensible language is the transition phase to the elected treatments (if any are available) and their implementation. [Fortunately, there exist no dirth of useful analyses of such bioethical problems, in addition to those in the extant bioethics literature like truth-telling (which I distinguish from "trust-telling") in the context of the doctor-patient relationship ([23], pp. 174–175), and the conceptual problems that surround the ethico-legal concepts of privacy, confidentiality, and informed consent.]

Logically and temporally prior to these bioethical issues, however, are those ethical issues that arise strictly within the complex cognitive process of differential diagnosis. For example, it is clear that diagnoses are *evaluative* since (1) there is ample evidence that patients value their bodily organs and "parts" differently, and (2) because physicians' diagnoses often signal actual or potential functional losses in contrast to the patient's understanding of his or her "ideal" functioning. Furthermore, since the process of diagnosis in present-day medicine necessitates the acquisition of highly subjective information along with objective, i.e., quantifiable, evidence from the patient, it presupposes certain costs to the patient, for the patient is required to evaluate what is is worth to him to undergo certain tests, perhaps painful ones, and other intrusions on his body. After all, there are costs in time, suffering, and dollars incurred as the physician proceeds toward a satisfactory level of diagnostic certainty. In this sense diagnosis is tied to what the patient (and physician) deem *worthy* of intervention, i.e., treatment. Thus, one can argue that the end of diagnosis is not merely epistemic – what the physician has to *know* to institute the most appropriate treatment – nor is it always immediately directed to interventions and treatments to alleviate pain, suffering and, perhaps, to forestall death, but often, and more importantly, aims at the establishment of a *trust relationship* between patient and physician, which is critical to establishing the limits of the diagnostic process (which I shall explore momentarily).

Given this fact, we can ask a series of related questions: How does a patient experience his body during the process of diagnosis, given the fact that the patient and the physician typically begin to communicate within the context set by the patient's lived body? What does it mean conceptually to be illness-embodied in the world? What implications does a diagnosis have with respect to our *integrated* sense of self and embodiment? That is, what role does the body as 'phenomenal field of meaning' play with regard to the ethics of diagnosis, and what bearing do diagnostic categories, labels, and classifications have on the physicians's and the patient's *acceptance* of the prognosis and, more importantly, on the limits of the patient's future life? How does the physician's actual assessment of the patient's body/organism establish an *ethical* obligation on the part of the physician?

IV. FROM INFIRMITY/PATHOLOGY TO SUBJECTIVITY

Consider, for example, José A. Mainetti's thesis: In order to acknowledge properly the essential "incarnation of the *mal*" and the anthropological category of "infirmitas" with respect to mankind, he asserts, it is necessary to transfer the concepts of health and illness from the technical-medical sphere to the "anthropo-philosophical". That is, it is misleading to conceive of the patient's body only "*sub specie pathologicae*", since this commits one to the view that infirmity, sickness, and various limitations are foreign to the essence of anthropos, whereas the tendency to become sick ("enfermabilidad") is, on Mainetti's view, natural to humanity and an essential aspect of being healthy [11]; i.e., being anthropos implies being imperfect. In sharp contrast to "pathologic ontologism", then, Mainetti calls for a "romantic" view of the body in health and sickness such that it is construed along a continuum from the normal to the pathological and back again, this being subject to but not solely restricted by quantification across a spectrum to gradual variations. The anthropo-philosophical account leads to the claim that "a normal person is anyone who has not been sufficiently investigated" ([14], p. 123). [5] In short, Mainetti maintains that abnormality and infirmity are of the essence of being human, and thus to describe the patient's body solely through medical discourse compels the "submission" of the patient to the order of *knowledge* and *power* where the patient is in danger of becoming a mere respondent to the command of medicine. This, he says in the spirit of Foucault, may come about "when medical classifications do not fully map onto generally experienced senses of pain and suffering", when, for example, there is ample evidence that the experienced pain and suffering may be both affirmed and rejected by patients because different patients reflect a range of values concerning their bodies.

V. INTEGRITY, BODILY INTEGRITY, AND TRUST-TELLING

For physicians, the anthropo-philosophical language used by Mainetti is not perhaps fully accessible, nor, however, does reliance on medical nosology capture or "fully map onto [a patient's] generally experienced senses of pain and suffering". Therefore, to be intelligible to physicians, it should prove useful to illustrate, by way of example, a case of a patient whose experience of illness reflects normative judgments concerning her personal/bodily integrity. Thus a compromise may be struck: the physician might find it useful both (1) to suspend the use of technical-medical language that may tacitly lead to taking power or control over the patient, and (2) to suspend all anthropo-philosophical language. Philosophers, Bernard Williams once remarked, repeatedly urge one to view the world *sub specie aeternitatis*, but for most human purposes that is not a good *species* to view it under. What remains, then, is the language of the patient.

Let us return to Barbara Webster's personal account of her chronic, long-suffered illness that remained undiagnosed for fourteen years, years during which "a young woman with poorly defined symptoms bounces from doctor to doctor, with no good evaluation of what is happening and no diagnosis" ([23], p. x). [I am reminded that such an experience was captured by a *New Yorker* cartoonist some years ago: A physician sits peering into the peep hole of his ophthalmoscope at his patient's fundus, pauses, then remarks: "My guess is, I don't know...."]

During the onset of her illness, Webster recalls that she experienced recurring urinary tract infections, problems with vision, gait, and balance (unable to walk in a straight line and at times impossible to walk due to inexplicable bouts of weakness in the legs), difficulty with talking and speech, sensory symptoms – pins and needles, numbness, tingling, "areas of solidity" – and extreme, "incredible" and "inchoate" fatigue that portended total immobility and inability to function. In those days, a series of misdiagnoses followed an equally long battery of tests, since MS was (and remains) a difficult disease to diagnose with accuracy: (1) nothing was organically wrong with her (it was all in her head, she was "a neurotic"), (2) she had a brain tumor, (3) she had a virus, (4) she had an unspecified nervous disorder, (5) she suffered an emotional disease, (6) she had a nervous disorder *caused by* emotional reasons – a combination of (4) and (5) – that required psychiatric treatment, (7) she was clinically depressed, and (8) she was sick for no good reason – a diagnosis which sounds strikingly like diagnosis (1) – nothing was organically wrong with her. The circle was complete!

Eventually, she was given the diagnosis of MS. With this *act* of diagnosis she began to be "immensely vindicated and affirmed" ([23], p. 23) as an individual, though her "social terrain" was dramatically altered: unlike persons with disease, those with handicaps neither recover nor die soon, they only change for the worse, usually quite slowly, are viewed as weak and dependent, and thus live in a rather special social limbo – frequently ignored or shunned even by health professionals – all of this becoming a continuing part of life.

Prior to the act of diagnosis, however – especially in this case, where fourteen years had passed – how did this patient experience her transformation in personal/bodily integrity? First, there was a capitulation: she *acquiesced* in the decisions others made for her by giving in to others' interpretation of her condition, others' definition of what she should do and be; furthermore, she experienced an *extreme powerlessness* and *hopelessness* in addition to extreme fatigue, and *she felt mistrusted* even by "the voices of medical authority" – yet all the time she "knew that there was something definitely [radically] wrong with my body but there was no one at that stage...who agreed with me" ([23], pp. 5,7,8,11).

Webster's subjective experience of her illness can be described as a transformation of her self image or sense of wholeness, her integrity, her being all of a piece. She remarks: "People did not trust me to know what was going on in my own body. They listened, instead, to the experts and the experts had dismissed me.... I was very reluctant to see a doctor for the most simple thing because I was afraid I would be seen as a hypochondriac or as simply hysterical" ([23], p. 6). In short, Webster found it very difficult to continue to believe in *her own interpretation* of what was happening to her. As she put it: "it was difficult to maintain a sense of intergrity in the face of all this disapproval..." though she soon realized that "my sense of integrity was in my own hands and I had to maintain it even if in the end I was the only one who believed in it" ([23], pp. 8, 10). This "sense of integrity", notwithstanding its more superficial psychological interpretation, serves as a transition to the moral point of view, and points beyond the psychological to the *ethical*. The sense of integrity signaled by Webster refers to her innermost agency and source of self, the sense of *autonomy* in which she hoped, indeed worked, to retain *self-control* and personal responsibility ([23], p.26) during the course of illness when she was subject to other people's judgments and interpretations of *her* experience. Here, it is important to distinguish autonomy as "self-control" from autonomy as "self-fulfillment", the latter illustrated by patients who demand that their desires, wishes, wants, and self interest all be served by the actions and efforts of others, including health care professionals. What Webster has so eloquently described is the incessant attack on her sense of integrity and self-control, the source of her choices. Indeed, this "attack" continued during the long and drawn out *process* of diagnosis that preceded the apparent power over reality given in the *act* of diagnosis, when she inadvertently saw her doctor's words in the diagnosis column – "multiple sclerosis".

In short, it is Webster's intentions and projects, her actions and interventions, that have great moral significance, and indeed it is not easy to think of any moral outlook which could get along without making some use of these intentions, projects, actions, and decision. Webster's decision to take up a range of projects and commitments, for example, and to refuse to be alienated from her self-identifying projects and decisions, by acknowledging and reaffirming her self as the source of her actions, projects, and convictions – "But Barbara you look so good." "Yes, but *I fell* so badly!" – all speak to her refusal to accept the rather detached and abstract judgments of others.

The immediate point of all this is not simply to stress the importance of this patient's deeply held convictions regarding her own integral experience of her body in illness, but to begin the process of reforming the so-called "fiduciary relation" between patient and physician, since the term 'fiduciary' does not quite capture each patient's moral right to a *trust-telling* relationship. How a patient phenomenologically responds to a physician's description of, for example, the patient's report of the fatigue of MS following the diagnosis, may not only seriously affect their relationship, but it may also affect the patient in such a way as to predispose her to experience an exacerbation of the symptom complex ([10], p. 1012.

Sometimes when I go for a walk, I start out fine and after a block or two have a slight limp. As I continue to walk, my left leg begins to drag and hit the sidewealk. And then, sometimes very suddenly, I simply have to stop – the legs stop working. The exercise of will has nothing to do with it ([23], p. 19).

At other times Webster told the physican:

My left leg was very weak and generally unreliable. My left arm felt quite dead and heavy, and it, too, was extremely weak. At the same time, I had a host of unpleasant sensations – tingling, numbness – all very hard to describe in a way that conveys much meaning to others ([23], p. 25).

Should the physician trust this report? Should these descriptions be addressed? Does the physician's inability to cure the MS have the force of logical entailment: ...therefore there is nothing more to learn, nor for me to say? The reports from patients like Webster make it clear that in such circumstances the patient's "very self is at issue and at stake in any real adjustment" ([23], p. 28), and the patient's condition is not reducible to the need for so-called "physical" adjustments. Trust-telling is also a process, one in which the physician acknowledges his belief in the correctness of the patient's descriptions, and attends to the impediments[6] to the patient's psychological acceptance as well as physical adjustments; paradoxically as it may sound at first, this may actually require the patient to negate hope and optimism, and abandon the search for total independence and the social recognition of his personhood.

At this moment, many are writing and arguing in support of the legal right to the Constitutional protection of patients and physicians to be free from intrusion by the state, absent a state interest sufficiently compelling to justify intervention into this private relationship. This line of argument is rooted in the premise that the interests of the patient and physician are the same, and that treatment decisions, following diagnoses, belong to patients and their physicians. In short, the doctor-patient relationship is continuously at risk of state interference and prohibition.

Trust-telling, however, goes beyond the law; it is grounded in the belief that there are decisions that are so personal, so private, and that so profoundly affect patients like Webster, that physicians (1) should be readily willing to accept the patient's descriptions as 'authentic', i.e., as correctly reporting his or her experiences (2) in order to make a conscientious diagnosis prior to treatment decisions and the provision of actual patient care. Trust-telling (and trust-listening), then, logically precede all legal means of protection or intrusion by government. It is the ethical nodal moment most pronounced when the clinical nodal moment – the chief complaint – is followed by the process of differential diagnosis.

VI. CONCLUDING WORD: DIAGNOSIS AND THE GOOD PHYSICIAN

Edmund L. Erde has called my attention to a "Sounding Board" essay, published in the August 20, 1981 issue of *The New England Journal of Medicine* where a physician, DeWitt Stetten, Jr., some fifteen years after receiving his ophthalmologist's diagnosis of macular degeneration (as well as reconfirmations from six additional and qualified ophthalmologists) felt compelled to complain about all seven physicians: they are interested in vision but have little interest in blindness, and therefore not one had "at any time suggested any devices that might be of assistance to me",... and none "mentioned any of the many ways in which I could stem the deterioration in the quality of my life" ([22],p. 458). Furthermore, he notes in passing, "as one loses vision one becomes disoriented not merely in space but also in time" ([22], p. 159).

The failure of contemporary physicians to earn the honor of "good physician" can be found in their attitudes toward handicapped (or potentially handicapped) patients. As internists are concerned with living patients, too often ignoring the diagnostic information to be gleaned from autopsies – possibly due to the association of cadavers with medical

120 STUART F. SPICKER

failure – so too a large number of diagnoses, like macular degeneration and multiple sclerosis, unduly suggest physician failure. As a consequence, the physician is in danger of missing an extraordinary opportunity to focus on and attend positively to each patient's self integrity and integral self image (more importantly his or her *lack* of it at the time of diagnosis), and thereby is far less likely to suggest strategies that may lead the patient to initiate the complex process of acceptance of an integrated new form of life so well described by Barbara Webster in *All of a Piece* [23]. This frequently missed clinical opportunity is far more serious than a physician's philosophically naive allegiance to a Cartesianism manifested in a bifurcated mind/body metaphysics; and it is surely not too high minded, either.

University of Connecticut
School of Medicine
Farmington, Connecticut, U.S.A.

ACKNOWLEDGEMENTS

* I am very grateful to Tom Bole III, Ph. D., Richard J. Castriotta, M.D., Roger Crisp, D. Phil., Eric T. Juengst, Ph.D., Julius Landwirth, M.D., Ian R. Lawson, M.D., and Robert U. Massey, M.D., for their time and interest in as well as for their suggestions and constructive criticisms of the initial manuscript; the remaining misexplications are, of course, entirely mine.

NOTES

1 As the clinician Henrik R. Wulff has remarked, "Some physicians seem to cherish the idea that clinical medicine is mainly applied physiology and biochemistry, and they regard the diagnostician as a detective who makes clever deductions from the clues offered by the patient... whereas the truth of the diagnosis is sometimes almost a matter of opinion" ([24], pp. 110–111).
2 See. Dr. H.R. Wulff's discussion. Chapter 11, 'Values and Clinical Science, ([24], pp. 160–176).
3 It is worth noting that in typical textbooks on diagnosis [6], like *Reasoning in Medicine: An Introduction to Clinical Inference*, authors typically provide a disclaimer: "Specialists in medical ethics have much of value to say about these issues [the values which clinical diagnoses presume], but their work goes beyond the scope of this book" ([1], p. 242). Yet these same authors frequently allude to "the risk of harm to patients", and note that it can not be eliminated but only minimized by knowledge ([1], pp. 258, 183). Such token remarks suggest that the ethical issues ingredient in diagnosis require still further attention. (For our discussion of ethical issues in *prognosis*, for which no space is available in this essay, see [21].)

[4] It would be misleading for the reader to conclude that with today's sophisticated advances in empirical tests to measure organic activities, death itself is now *unambiguously diagnosed* by physicians. Indeed, the contemporary debate on the nature of the criteria by which to diagnose personal death, which involves brain and CNS evidence, reveals at least two schools of thought: (1) the neocortical camp that argues that a patient is dead once the *entire neocortex* is necrotic, and (2) the *whole brain* school that requires that the entire brain, including the brain stem, be necrotic (isolectric) in order to declare the patient dead, i.e., make the certain diagnosis of death (see [25]).

[5] Unfortunately Mainetti misconstrues the meaning of "Proposition 3" that he borrows from E. Murphy; for Murphy is making the very opposite point for his purposes: He remarks that "Proposition 3" "is a goldmind of rhetoric . . . . It illustrates paraprosdokian, hypallage, and syllepsis with a subtle enthymeme". Indeed, the proposition is "patently absurd" (see Murphy's full account [14], pp. 123−124).

[6] It should be pointed out that an increasing number of physicians have requested further training in psychological counselling germane to their work with MS patients and their problems. (See Alexander Burnfield's personal and thorough account of his MS in 'Doctor-Patient Dilemmas in Multiple Sclerosis', that anticipates a number of critical observations and insights made by Barbara Webster [2].)

BIBLIOGRAPHY

1. Albert, D. A., Munson, R., and Resnik, M.D.: 1988, *Reasoning in Medicine: An Introduction to Clinical Inference* (The Johns Hopkins Series in Contemporary Medicine and Public Health), Johns Hopkins University Press, Baltimore Maryland.
2. Burnfield, A.: 1984, 'Doctor-Patient Dilemmas in Multiple Sclerosis', *Journal of Medical Ethics* **10** (1), 21− 26.
3. Engelhardt, H.T., Jr.,: 1980, 'Ethical Issues in Diagnosis', *Metamedicine* **1**(1), 39− 50.
4. Fossel, E.T. *et al.*: 1986, 'Detection of Malignant Tumors: Watersuppressed Proton Nuclear Magnetic Resonance Spectroscopy of Plasma', *New England Journal of Medicine* **315**, 1369–1376.
5. Griffith, R.M.: 1970, 'Anthropodology: Man A-Foot', in *The Philosophy of the Body*, S.F. Spicker (ed.), Quadrangle Books, Chicago, Illinois, pp. 273–292.
6. Hurst, J. W. (ed.):1989, *Criteria for Diagnosis*, Butterworths, Boston, Massachusetts.
7. Kassirer, J.: 1989, 'Our Stubborn Quest for Diagnostic Certainty: A Cause of Excessive Testing', *New England Journal of Medicine* **320** (22), 1489–1491.
8. Laor, N., and Agassi, J.: 1991, *Diagnosis: Philosophical and Medical Perspectives*, Kluwer Academic Publishers, Dordrecht, The Netherlands, especially Chapter IV, "The Ethics of Diagnostic Systems", pp. 67–90.
9. Leder, D.: 1992, 'The Experience of Pain and its Clinical Implications', in this volume, pp. 95–105.
10. Love, S. M., Gelman, R.S., and Silen, W.: 1982, [Sounding Board] 'Fibrocystic "Disease" of the Breast − A Nondisease?', *New England Journal of Medicine* **307** (16), 1010–1014.
11. Mainetti, J.A.: 1992, 'Embodiment, Pathology and Diagnosis', in this volume, pp. 79–93.

12. McCullough, L. B., and Christianson, C.E.: 1981, 'Ethical Dimensions of Diagnosis: A Case Study and Analysis', *Metamedicine* **2**(2), 129–143.

13. Moskowitz, R.: 1983, 'Vague, Long-Term Diagnosis: The "Nocebo" Effect', *Journal of the American Institute of Homeopathy* **76**, 26–28.

14. Murphy, E.A.: 1976, *The Logic of Medicine*, Johns Hopkins University Press, Baltimore, Maryland.

15. Nelkin, D., and Tancredi, L.: 1989, *Dangerous Diagnostics: The Social Power of Biological Information*, Basic Books, New York.

16. Rosenberg, M.: 1990, [Letter] 'History and Examination Should Precede Tests', *Journal of the American Medical Association* **263** (12) 1632.

17. Schaffner, K.F. (ed.): 1985, *Logic of Discovery and Diagnosis in Medicine*, University of California Press, Berkeley/Los Angeles, California.

18. Shaw, A.: 1987, 'Ethical Dilemmas in the Early Detection of Malignancy by NMR Spectroscopy of Plasma', *New England Journal of Medicine*, **317** (21), 1353.

19. Sox, H. C., Blatt, M.A., Higgins, M.C., and Marton, K.I.: 1988, *Medical Decision Making*, Butterworths, Boston, Massachussetts.

20. Spicker, S.F.(ed.): 1970, The Philosophy of the Body, Quadrangle Books, Chicago, Illinois, (Intro.) pp.3–23.

21. Spicker, S.F., and Raye, J.R.: 1981, 'The Bearing of Prognosis on the Ethics of Medicine: Congenital Anomalies, the Social Context and the Law', in S.F. Spicker, J.M. Healey, Jr., and H. T. Engelhardt, Jr. (eds.) *The Law-Medicine Relation: A Philosophical Exploration*, D. Reidel Publishing Co., Dordrecht, Holland / Boston, U.S.A., pp. 189–216.

22. Stetten, D.: 1981, 'Sounding Board', *The New England Journal of Medicine* **305** (8), 458–460.

23. Webster, B.D.: 1989, *All of a Piece: A Life with Multiple Sclerosis*, Johns Hopkins University Press, Baltimore, Maryland.

24. Wulff, H.R.: 1981, *Rational Diagnosis and Treatment: An Introduction to Clinical Decision-Making*, 2nd ed., Blackwell Scientific Publications, Oxford, England.

25. Zaner, R.M. (ed.): 1988, *Death: Beyond Whole-Brain Criteria*, (Vol. **31**, *Philosophy and Medicine* series) Kluwer Academic Publishers, Dordrecht, The Netherlands.

BODILY NORMS AND THE ETHICS OF DIAGNOSIS

What are the ethical obligations of the diagnostician? A patient presents herself with certain complaints, and these complaints reflect deficiencies she perceives in somatic (or psychosomatic) form or function. These deficiencies only make sense in terms of the values she places on her body. The prospect of mastectomy will mean different things for the aged nun than for the young female professional still planning on marriage and family, and the different values and designs each has for her body ought to be appreciated by the physician in diagnosis. As Mainetti points out, the patient's views of what calls for medical intervention may differ from the physician's [3]. In case there is a difference, whose values ought to have priority, and why? (This question is ethical, not primarily legal.) Can the diagnosing physician claim priority for the norms of good health? Or ought these norms be subjected to the patient's informed desires?

Spicker [4] and Leder [2] suggest that the latter alternative is the correct one. Spicker contends that the diagnostician ought to take the patient's reports of her experience of suffering as "'authentic', i.e., as correctly reporting his or her experiences", in order to make the correct diagnosis ([4], p. 119). Leder observes that the physician ought to diagnose and try to treat the suffering patient, not just her diseased body. Both Spicker and Leder imply that the diagnostician ought to try to appreciate and to be guided by the goal of treating the suffering patient as well as by the goal of treating the underlying disease. But what is the relationship between these goals, and what should their priority be relative to each other? The physician, of course, wants a cooperative patient who feels well served, not one who might, for example, initiate a malpractice suit. Even so, the same questions remain that arise about Mainetti: Ought the physician to take any more than a prudential account of the patient's goal, so that he attempts to treat the concerns of the patient insofar as they guide him in treating the disease? Or ought the physician to allow the patient's concerns to guide treatment, so long as she is competent and takes account of the physician's informed medical opinion?

J. L. Peset and D. Gracia (eds.), The Ethics of Diagnosis, 123–127.
© 1992 by Kluwer Academic Publishers. Printed in the Netherlands.

What is the sort of problem with which the diagnostician is supposed to be concerned? If there is some canonical notion of somatic form or structure, this notion will dictate appropriate boundaries to medical diagnosis and treatment. If there is no such notion, on the other hand, diagnosis will seek pathoanatomical and/or psychopathological causal explanations for some perceived disease, disturbance, or difficulty, and will do so in order to see what can be done to alleviate that problem; but what counts as such a problem will depend upon how the patient values her body. I shall argue (1) that there is no canonical notion of somatic form or structure, (2) that what counts as a problem for diagnosis depends upon the particular values the patient invests in her body, and (3) that in consequence there are no general values for the diagnositican, save to get the competent patient's free and informed consent to pursue what the dianostician judges the wisest course.

(1) The question of whether or not there is a canonical notion of somatic form or structure implies the question of whether or not there is a canon deviations from which should guide medical diagnosis. If there is such a canonical notion, there is a canon for medical diagnosis. Then the ethics of diagnosis would be dictated by the value of conforming to that canon, the value to be placed upon the patient's informed consent in pursuing that conformity, and the weight of each of these values relative to the other. If there is no such canon, then the diagnostician cannot assume that what is medically indicated, even if it is clearly medically indicated, represents a medical good that should be normative for the patient.

Diseases, disorders and disturbances are deviations from physiological or psychological norms. These norms, however, are the results of individuals' judgments about proper form and function and freedom from pain. Such judgments are aesthetic as well as functional. What appropriate form and function are will differ vastly, e.g., between elderly nun and young unmarried woman, or between the marathoner and the weightlifter. They are not at all agent-neutral, I contend, but agent-centered – although in these cases the agent is the patient.

One may be willing to concede this contention with respect to the aesthetic, but nonetheless object that there are certain functions that are relatively objective, and that the impairment or limitation of these functions are objective disvalues because they impair or limit any human being. Such negative values surely reflect an objective canon of values to which the diagnostician should adhere. There are descriptive judgments about

diseases, e.g., cancer or schizophrenia, which are relatively free of particular individual or cultural values. They are judgments about states of
affairs that inhibit human beings' well ordered functioning, and which
the physician ought to try to diagnose and treat.

This objection assumes that it makes sense to talk about functions
apart from the standards which define good or ill functioning. Nature,
however, has no norms, only statistically average behavior. It is not
typified in fixed Aristotelian essences and final causes, and does not, or
does not obviously, reflect the design of a benevolent Creator. Moreover,
functions, at least insofar as their limitation or impairment determines
someone as ill, only make sense in terms of the individual human beings
and societies living in particular environments. Diseases such as cancer
are relatively objective disvalues, because they cause a degree of pain
and circumscription of goals that practically all human beings have good
reasons to disvalue. Such dysfunction and disvalue, it should be noted,
are dictated by particular human goals in particular cirsumstances, not by
species-wide functions of adaptability to the environment. Even a relatively non-controversial diagnosis of cancer can be rejected, in the sense
that the patient, if competent and acting in a way that does not harm
others, can refuse clearly indicated treatment and deny the cancer's disvalue by leaving it untreated. This possibility shows that the problems
that bring patients to the diagnostic clinic ultimately make sense in terms
of particular invidual' valuations of somatic form and function within the
particular environs in which they find themselves.

In fine, there is no objective standard of somatic values that should
override those that the competent and informed patient invests in her
body. This point is an application of the general point that one cannot
establish a canonical standard of value that morally justifies him in
coercing some other moral agent who is not committed to that standard
([1], pp. 32–37). One cannot, for example, show that a particular ranking
of the relative values of longevity of life, freedom from pain, freedom
from dysfunction, or freedom from malformation, is normative for
someone else without presupposing a standard reference which ranks
these values, and from which the other person may dissent. Nor can one
argue from the consequences of submitting or not submitting to some
diagnostic procedure without presuming a canonical standard of how to
rank consequences for long life relative to those for freedom from pain,
etc.; and again, the other person can dissent from the standard. There is,
then, no rationally canonical standard of values that justifies one person

imposing it upon an unconsenting innocent other. A fortiori, there is no standard of values that would justify imposing certain diagnostic procedures upon patients.

(2) It follows that there is no canonical standard of the values, moral or non-moral, of diagnosis. There is no account of the significance of the subject of diagnosis, the patient's body, apart from the projects and values of the particular person whose body it is. In the aftermath of an accident involving essentially similar leg injuries, the marathoner would probably be willing to undergo riskier procedures than the monk to regain function. What counts as a clinical problem, then, depends upon the patient's particular projects and values and how his or her body figures into them.

(3) These values should obviously inform the physician, the more so to the extent to which the appropriate diagnosis and treatment of the patient's clinical problem is or remains unclear. The diagnostician will have to ask what the costs are for reaching the decision that a particular diagnosis is true, and what the costs are for treating on the basis of that diagnosis. The costs of a radical mastectomy for breast cancer are obviously very severe for a young woman desiring marriage and family. The costs, however, of delaying treatment for confirmatory diagnoses may also be very severe. The diagnostician must weigh the costs of ascertaining the additional knowledge against the costs of delaying treatment.

In such a weighing process, the interests and values of the patient play the pre-eminent role. This is not simply because of the uncertainties involved in making diagnoses and prescribing treatments. It is also because, in the absence of an objective standard of such values, the patient must be presumed to be the one most familiar with those values and interests in her body that define what would be best in particular circumstances for her. The physician must in consequence gain moral authority to diagnose in the same way as the physician gets moral authority to treat: not simply by getting the competent patient's free and informed consent to pursue the course of action informed medical judgment deems best, but also by making sure that judgment is directed by that patient's values.

If the ethics of diagnosis ought to take seriously the values of the suffering patient, as Mainetti suggests, the appropriate grounds for doing so may be more radical than those suggested by Leder and Spicker. The diagnostician should not simply be guided by the goal of treating the patient's suffering as well as the underlying disease. Rather, the diagnostician's goal in trying to diagnose the underlying disease should ultimately serve the patient's competently expressed somatic values and interest.

University of Oklahoma Health Sciences Center
Oklahoma City, Oklahoma, U.S.A

BIBLIOGRAPHY

1. Engelhardt, H.T., Jr.: 1986, *The Foundations of Bioethics*, Oxford University Press, New York.
2. Leder, D.: 1992, 'The Experience of Pain and its Clinical Implications', *in this volume*, pp. 95–105.
3. Mainetti, J.A.: 1992, 'Embodiment, Pathology and Diagnosis', *in this volume*, pp. 79–93.
4. Spicker, S.F.: 1992 'Ethics in Diagnosis: Bodily Integrity, Trust-telling, and the Good Physician', *in this volume*, pp. 107–122.

SECTION III

THE SOCIO-CULTURAL DIMENSION OF
MEDICAL KNOWLEDGE

MARX WARTOFSKY

THE SOCIAL PRESUPPOSITIONS OF
MEDICAL KNOWLEDGE

The title of this volume, *The Ethics of Diagnosis*, suggests a problem that is directly related to the subject of my essay. The problem, in short, is this: on the one hand, diagnosis is a case of scientific judgment, a cognitive procedure that aims at discovering the answer to a factual question, i.e., "What disease or illness is the patient suffering from?" or "What is the cause (or what are the causes) of the symptoms, signs, and complaints that the patient presents?". On the other hand, if one poses the question about the *ethics* of diagnosis, then presumably, this concerns normative or valuative considerations that go beyond what are usually regarded as purely scientific or cognitive contexts. Such considerations are presumably not simply factual but concern what is ethically right or wrong. I would suppose that such ethical or normative considerations bear upon the well-being of the patient, the moral responsibility of the physician, or on his or her honesty or integrity; or upon considerations of the values involved in the diagnostic process itself, with respect to costs and benefits, or the risks involved in the use of certain diagnostic tests, etc.

In the traditional way of dealing with such questions, matters of fact are taken to be subject to objective or scientific determination, within the limits of the fallibility of human knowledge and the constraints of human ignorance. Ethical questions, by contrast, are taken to be normative or valuative, and not factual or descriptive; and therefore to be determined on grounds other than scientific ones. The problem that this classical dichotomy between facts and values poses is the following: Is the diagnostic process divided into two discrete parts, one of which is "objective", "scientific", "matter of fact", and the other of which is "subjective", "non-scientific", valuative"? Does the diagnostician first establish a factual judgment and then consider what ought to be done about it, in ethical terms? David Hume proposed the clearest view of such a procedure, holding that reason or scientific judgment laid out all the facts before us, but that it was sentiment or "an active feeling" that decided on what ought to be done in the light of these matters of fact.

This frames the issues that I want to address in this essay concerning

131

J. L. Peset and D. Gracia (eds.), The Ethics of Diagnosis, 131–151.
© 1992 *by Kluwer Academic Publishers. Printed in the Netherlands.*

the social presuppositions of medical knowledge. The question that I raise is the following: Can the distinction between the factual and the descriptive, on the one hand, and the valuative or the normative, on the other, be maintained concerning medical knowledge? I will argue that it cannot; that medical knowledge is *essentially* normative, and that the cognitive or scientific content of medicine cannot be understood apart from this normative character. Further, I will argue that this normative character of medical knowledge derives from the *social* nature of medical knowledge. The particular context of this sociality that I want to address here is the responsibility that the medical practitioner has in the acquisition, the use, and the preservation and transmission of medical knowledge. This bears directly, as I hope to show, upon the question of *rights*, in at least two respects: the rights of the physician in the practice of medicine, and the rights of the patient with respect to the availability and the quality of medical care and treatment.

My argument will proceed in three parts: first, I will propose an epistemological thesis concerning the nature of medical knowledge; second, I will attempt to show that the question of rights – of the right to practice medicine, and the right to medical treatment – needs to be resolved in the context of the social nature of medical knowledge itself, that is, that the social-theoretical issues in medicine are essentially linked to an adequate epistemological theory of medical knowledge. Finally, I want to address the specific question of diagnosis in this context, namely, in terms of the social relations between physician and patient in the diagnostic process itself. Therefore, in considering the epistemological issues involved in diagnostic procedure, as a case of scientific judgment, I will argue that the science of diagnosis cannot be reconstructed or properly understood, either epistemologically or methodologically, apart from the contexts of a normative social theory of medical practice.

I. THE DISTINCTIVE CHARACTER OF MEDICAL KNOWLEDGE FROM AN EPISTEMOLOGICAL POINT OF VIEW

Medical knowledge, like all knowledge, is a social artifiact. That is to say, it is the product of an activity that is engaged in by human beings as a common or social activity, and in response to a social need. Granted, it is only individuals who acquire, use, and transmit such knowledge, but they do so within the forms of a social and linguistic community and within specific forms of social practice and organization, however these may differ from one place to the next, or from one time to another. Now these are apparently vacuous and harmless platitudes, and it would be redundant to repeat them if it were not the case that they tend to be over-looked or forgotten in many theories of knowledge. Let me make this point a bit clearer: in traditional philosophical epistemologies, whether of the rationalist or empiricist sort, as these have been developed from the seventeenth and eighteenth centuries, knowledge has been characterized as that which is acquired by an individual, as a consequence either of his experience or of his cognitive structure. The universal features of such knowledge acquisition by individuals is based on the presupposition (or the argument) that all individuals are the same, either in capacity or structure, e.g., that they are all rational, or that they all have the same sensory or perceptual capacities. The sociality of knowledge is therefore at least tacitly affirmed, insofar as all human being are considered to share in these universal features of reason or sense-perception. But this is a presupposed and *passive* sociality. What I am proposing instead is that the social character of knowledge has to be understood as a function of the social interaction among individuals, and more specifically, of the concrete forms of social interaction that are typical of a given society or a given historical period. One may speak of the biologically *a priori* con-ditions for the acquisition and use of such knowledge – e.g., a certain development of brain and neural system, or the organs for speech and other forms of symbolic communication – and one may also speak of general or trans-historical *socially a priori* conditions for such knowl-edge, e.g., some pre-existing form of social life, a linguistic community, the growth of techniques of production or of social organization, etc. But this gives us only the conditions for the possibility of knowledge, or of the acquisition of knowledge as a human activity. It does not yet specify how knowledge is generated, or why it is generated; nor does it give us, as yet, any concrete or specific understanding of this knowledge itself.

Every particular form of knowledge therefore has to be specified with respect to something more than these conditions. It has to be specified with respect to the needs that this knowledge serves, and with respect to the recognition of these needs that is expressed in the conscious purposes to which this cognitive acquisition is addressed. In short, both in its modes of acquisition and in its uses, human knowledge is teleological; more precisely, it is a teleological artifact. It is acquired and used for the sake of some end. But this is not a narrowly pragmatic or utilitarian view of knowledge, for the ends for the sake of which such knowledge may be acquired may be characterized in a variety of ways. For example, we may say that knowledge has, as a general aim, the discovery of truth; or, on another view, we may say that knowledge is pursued simply for the pleasure or satisfaction that the activity of cognition affords us, or as a playful activity for the sake of aesthetic satisfaction. Each of these ends may be seen as the conscious recognition of a particular human need. In the complex economy of human life, the need for play, or for truth, or for freedom is as real a need as the need for food, for companionship, or for good health. I am not ready to propose a hierarchy of ends in which some are more intrinsic or higher than others, since this is not at issue here. Rather, what is at issue is the particular characterization of medical knowledge in terms of its ends as a certain kind of teleological artifact. Thus, what distinguishes medical knowledge, what individuates it as *medical*, rather than as biological, historical, or moral knowledge, is the distinctive ends that it serves. As such, medical knowledge may include biological, or historical or moral components; but these are then ordered with respect to the ends that medicine serves.

How, then, are we to determine what the ends of medicine are? We could simply stipulate what these ends *should* be, as normative proposals, or we could empirically describe what medicine does, as a matter of fact, and then try to interpret its intrinsic ends in terms of its deeds. Rather than describing or prescribing what the ends of medicine are, however, there is a more direct and perspicuous way to discover them. That is to examine what the norms are of the practice of medicine itself, to see what the profession or the practice of medicine specifies as its own ends in the course of its practice. We will find that it is not simply a question of describing what physicians do, but rather of what they take themselves to be doing, as well as of what patients take them to be doing, and what the society at large takes them to be doing. What I want to avoid here is some prescriptive assertion of what *I* think medicine ought to have as its aims, though this would certainly be a normative characterization of medical knowledge. On the other hand, I do not think that the norms of medical practice or of medical knowledge are simply the subject for a descriptive sociology of knowledge; rather, they are also subject for a critical reflection on their adequacy or their coherence. In this essay, however, I do not propose to undertake this task. It will be sufficient for my purposes here to establish in an informal and *ad hoc* way that medicine does in fact have such norms , that it is a teleological practice, and one thing more: that the norms change historically, because they reflect different social modes of constituting the practice of medicine and different conceptions of the role and value of medicine.

In a general way, we may propose that the ends medicine has undertaken to meet throughout its history are the care or cure of the ill, the treatment of disease, the alleviation of suffering caused by illness, the prevention of untimely death, and more generally, the prevention of disease, or of its spread. Historically, there have been varying emphases and interpretations of these various ends. But whether the emphasis is on treatment and cure, alleviation and maintenance, or on prevention, the phenomenon with which medicine has dealt has been human weal and woe as it pertains to bodily function, to the patient's ability to function in ways required by social and personal life, and to the patient's complaints and feelings of pain, illness, or incapacity. There is, or course, another large area that has been of differential concern to medicine, in various cultures and even within a given culture, and that is the concern with the promotion of positive health, as more than the absence of disease. The relation of this aspect to curative and merely preventive medicine is a complex question, which I will only note here. What is clear from all this is that medicine is defined with respect to ends that concern the well-being and ill-being of individuals, and with the effects of these states of being on social life in general (and specifically, on the carrying on of social production, of family life, and the maintenance of viable social organization or institutional life).

The ends of medicine are therefore the satisfaction of a specific range of needs, and the perception and recognition of these needs is what generates the values of medicine. But these needs are not merely the needs of isolated individuals, and neither are the values. All needs are ultimately needs of individuals; but the needs of a given individual, which when recognized may come to generate values for that individual – i.e., the value of whatever would satisfy such needs – are not in themselves enough to determine the ends of medical practice. One may imagine a powerful individual whose needs, as such, may dictate medical practice, as for example, we may imagine that the needs of the Egyptian Pharaohs or of kings or millionaires may have come to dictate the practice either of specific court-physicians or even of the profession as a whole. But this is to lose sight of the fact that such individuals represent in their very power a certain set of social relations. For example, the Pharaoh is not simply some abstract individual, but rather the embodiment of Egypt itself, or of the Nile. The life of the Pharaoh, and his death (involving the development of anatomical and surgical techniques, embalming, etc.) are themselves taken to be not only symbolically, but actually, social phenomena embodied in a personality. Similarly, the case can be made for kings and millionaires, whose power, role, and wealth are expressions of social authority and of a certain form of social relations. So it is never individuals *as such* who define the ends of medicine, or its varying norms, but always individuals in a certain mode of social relations.

Nor is it only the ends of medicine that are social in this sense. So too are its means. Medical knowledge is the collective knowledge of the medical community, or of the medical guild, accumulated, transmitted, preserved, and transformed in the course of a practice marked by distinctively social forms of organization, with its rules, its mores, its self-regulating authority, its social legitimation, its ideology. To be a physician at all is to be a member of a certain community, the means of whose practice is socially controlled, acquired, and transmitted. With the division of labor, and with the genesis of a professional class of physicians or of surgeons, and with the further development of specialization, the practice of medicine becomes more and more socialized, as the fragmentation of functions and specialties demands greater cooperation among the specialities. The conclusion from all this is that medicine is a social practice, that it is defined by its ends, and that these ends are expressed in the norms of this practice; further, that these norms express the values of medicine, and that these values, in turn, are generated by the social recognition of needs. That is not to say that medicine is simply a practice that meets human needs, but rather that it organizes itself around those needs that come to be socially recognized as needs that ought to be met. A critical social history of medicine would point out that the profession of medicine, or the social context in which the profession was practiced, very often failed to recognize real needs, and that the values of the profession have been, in many cases, either wrongly or too narrowly conceived; or that the profession has served only some at the expense of others, and in this way has served to reinforce existing forms of inequality or of social domination. Medicine as a social practice is therefore not yet medicine as an egalitarian or a socialist practice, though the principle of equal rights to medical care has been voiced as an abstract principle throughout the history of medicine.

Thus far I have been using the terms "medicine", "medical practice", and "medical knowledge" almost interchangeably. Let me now justify this usage. By "medical knowledge" (ostensibly the subject of this paper), I mean not only the theoretical or didactically transmissible knowledge that consists in knowing that something is the case, or knowing what is true. This is only one component of what I would characterize more broadly as knowledge. In fact, I would propose that the term "knowledge" here be used to include three distinguishable components: *first*, theoretical knowledge, or what has come to be characterized philosophically as "propositional knowledge", or knowledge that something is the case; *second*, technical or practical knowledge, or what has come to be philosophically characterized as "knowing how", by contrast to "knowing that", and *third*, what I would call ideology, or knowledge of the norms, values, or acceptable modes of procedure or of behavior. The first sort of knowledge constitutes what is contained in the didactic curriculum of medical education, namely, the basic sciences, and in all the more special medical sciences, e.g., anatomy, physiology, pathology, pharmacology, etc. The second obviously pertains to the clinical component in the medical curriculum, but is inseparable from the first in the sense that it is a matter of learning how to apply theoretical knowledge in specific cases; but it also includes the development of those capacities, skills, and techniques required for diagnostic and therapeutic practice. These include not only the more or less "manual" technical skills, in learning to use instruments, but also the judgmental skills, or casuistical skills required to bring one's background knowledge and one's reasoning and observational powers to bear upon individual cases. The third kind of knowledge is not so much taught as absorbed in the course of medical education and medical practice (though more recently, it is coming to be a more explicit part of the curriculum). This includes the acquisition of the value-system, of the norms of good practice, of the manners and mores of the profession, as well as the sense of responsibility for one's decisions and actions, and the acceptance of the authority-relations that characterize the profession. Insofar as it deals explicit with criteria for judgment, with respect to what it is right or wrong to do, and insofar as it explicity concerns *rights*, *duties*, *culpability*, and *obligations*, this kind of knowledge includes medical ethics, social theory, and the legal contexts of medical practice.

On such a broad view of what constitutes medical knowledge, the practice of medicine is in effect identical with the domain of medical knowledge. But one additional argument is needed here: In general, there has been a division between "theoretical" and "practical" knowledge, introduced in a systematic way by Aristotle, which distinguishes these two kinds of knowledge with respect to the differences in their ends. Theoretical knowledge aims at truth for its own sake; practical knowledge aims at the good of its object, or at what will help to realize its object's fullest potentialities. The objects of practical knowledge are either things made (*factibilia*) or actions done (*agibilia*). The second kind of object is done for the sake of other persons. Thus, practical knowledge of the second sort has to do with what concerns the good of persons, insofar as they act with respect to each other. Its domain is therefore the domain of social life, or of social interaction, or what Aristotle defined as *ethics*. Now, if a component of medical knowledge is theoretical, then part of medicine would seem to have as its ends, and therefore as the norm of its practice, truth for its own sake. And yet, the medical sciences proper, i.e., those that do not have to do directly with the treatment of patients but rather with the discovery of scientific truths about the life-functions of organisms, both physiological and pathological, are often taken to be sciences in the service of medicine: the theory is for the sake of the practice, even where it is pursued autonomously without regard to any specific goods. On this I would argue (as others have done) that the very conditions of medical-scientific research, whether at the level of whole organisms, or at the tissue, cell, or biochemical level, are essentially involved with the conceptions of normal and abnormal function, with viability of the organism under different conditions, and therefore with a normative context that carries over into the precincts of "pure research" the same basic norms which inform clinical practice. However dispassionate the inquiry into basic biological or biomedical processes may be in the so-called medical "sciences" (by contrast with clinical practice), it is never simply an account of such processes abstracted from their relation to the life processes of a whole organism, and therefore it is never an account wholly abstracted from the considerations of normal function. When the organism is human, then the very conception of "normal" function can no longer be taken in a purely biological way, but involves considerations of the social mode of being that characterizes human life. On my view, then, medical knowledge, in its theoretical, as well as in its practical and ideological components, is essentially normative, that is, it has to do with the *good* of its objects, where this "good" is at least minimally defined with respect to normal function.

It is a notorious problem in the philosophy of medicine, as in other areas in philosophy that deal with what is normative, that "normal" is an exceptionally slippery term, and that to identify "good" with "normal" in effect begs the question. I do not intend to enter into this discussion here, but rather to sidestep it by what I think is a warranted evasion for my present purposes. Let me propose that the burden of the term "normative" as I use it be borne by the concepts "better" and "worse". Thus, "normative" in the sense of conforming to a norm is not in itself enough to characterize the valuative content of the concept that I want to retain. The norm itself has to be one that connotes a better state than if it were not attained. In this sense, whatever the particular interpretation of the norms of medical practice may be, it is generally clear that the conformity to these norms is regarded as the mark of a better state, and that failure to attain the norms is regarded as the mark of a worse state. Thus, if health, or the absence of disease, or normal function is regarded as the end of medicine, and the norms of the practice embody this as a value to be achieved, then such conditions are not merely descriptions of a state of the organism, nor merely normative in the thin sense of a statistically "normal distribution in a population". Rather, these conditions are "better" when present, and their absence is therefore "worse". The idea that there is some "value-free" description of states of an organism may be a useful abstraction in some contexts; but my overall argument is that such an abstraction fails to capture something that is *factually* true about organisms in general, and in a stronger sense true of human beings: namely, that "better" and "worse" are accurate descriptions of an entity whose mode of being is essentially normative. The difficulty in this argument thus far, is that "better" and "worse" are, from the point of view taken here, constituted as judgments *within* a practice, or within a mode of knowledge – in this case, medical knowledge – and I have stated that an account of the norms of a practice is most accurate in terms not simply of what the practitioners "do" or how they "act" in some allegedly context-free description, but rather of what they take themselves to be doing, or how they are understood to be acting by others as well. In short, the price I seem to have to pay for reconstructing the concept of medical knowledge as a social construct is to be caught within the historical or cultural relativism of varying determinations of what is "better" or "worse". Let me suggest that the way out of this problem is not a reversion to some transcendental ground for the *Good*, with respect to which we can then measure medicine's own norms of what is better or worse, but rather an historical reconstruction of the conditions and arguments that led to various definitions or characterizations of the ends of medical practice, *and* an attempt to discover whether there are some trans-historical and trans-cultural norms that persist. Such an historical reconstruction of the norms of medical practice and the variations and invariances that this history of norms exhibits does not in itself suffice as a critical or Archimedean point upon which to ground our own contemporary view of which norms are better, worse, or more or less adequate than others. Yet, it provides some protection against dogmatism or insularity with respect to contemporary norms, and suggests to us at least that our present norms are themselves historically and culturally conditioned. It therefore encourages a fallibilism and an open-mindedness that contributes to a critical and self-critical understanding.

Contemporary practice, of course, has a certain priority and legitimacy in this regard, since we generally hold that the norms that contemporary medicine has arrived at are "better" in some sense than those that they have replaced. This view is supported by two arguments at least: first, that the older norms have been rejected on some rational or experiential grounds, that is, they have been found wanting or inadequate as a result of criticism and new experience, or in view of new discoveries; second, that the newer norms that have replaced the rejected ones have exhibited their superiority in the empirical consequences of having adopted them, for example, people live longer, or realize their life-potentialities more fully; or states that were once taken to be "normal" or necessary burdens of life are now recognized as diseased or abnormal states, and therefore, medical practice comes to take them as part of its concern. Both of these arguments are plausible, but not yet definitive, since there may be consequences of older norms that, had they been realized, would have led to improvements in life-conditions, or in conditions of research and medical practice; and these prospective advantages have been lost in abandoning these older norms.

What I am suggesting is a certain analogy between approaches to the question of progress in science, or to the growth of knowledge in the acceptance and rejection of scientific theories, and the question of progress in normative contexts, in the history of the acceptance and rejection of norms. An adequate treatment of this question is part of the task of an inquiry into the logic and history of normative concepts, i.e., of concepts such as "good", "better", "worse", "ought", "need", etc., and more specifically, of such normative concepts as "health", "disease", "normal function", "optimal function", "illness", and "incapacity", in medicine.

What I am suggesting beyond this, however, is that any theory of medical knowledge is incomplete without just such an analysis, once it is recognized that medical knowledge is essentially normative, or concerns the knowledge of what one may term "normative facts". It is also essentially normative methodologically, in the sense that the methodological rule of thumb that operates in both medical theory and practice is that the practitioner or theorist is responsible for "doing the best he or she can do". Responsibility and culpability in medicine would make no sense unless there were at least some such concept of "doing one's best", for then, failing to do so could not be defined. Practically speaking, "doing the best one can" presupposes some definition of the state of the art, i.e., of the state of contemporary theory and practice, or of medical knowledge that determines what the *competent* physician ought to know, or ought to know how to do. Medical responsibility therefore entails some category that is at the same time epistemological and normative; for it is not merely a case of doing the best one can with respect to the knowledge one does in fact have, but rather with respect to the knowledge one is expected to have, or indeed, *ought* to have. But this "ought" with respect to medical knowledge brings me to the second part of this paper, namely to the question of rights. Therefore, let me summarize the argument thus far:

The arim of the foregoing account of medical knowledge was to establish a) that medicine is a social practice and that medical knowledge is socially constituted; b) that medicine is a teleological practice. i.e., that it is defined by its ends, and that it is these ends that distinguish medicine from other practices, including those to which medicine is closely related; c) that the term "medical knowledge" is not limited to "propositional knowledge" alone, but that it includes in its denotation practical or technical knowledge as well as knowledge of norms, or an ideological component, in addition to the theoretical or propositional component; and the therefore medicine as a cognitive practice is coextensive with medical knowledge taken in this broad sense; and d) that medical knowledge is essentially normative, and therefore that the division into a "factual" and a "normative" component violates not only the content of medical theory and practice, but also fails to capture the very nature of the object of medicine, the functioning organism, or more specifically, the functioning human being, as an essentially normative being.

There is a small quandary about this last point that needs to be cleared up, by way of transition to the discussion or rights. It is this: I stated at the outset that the normative character of medical knowledge derives from its social character, i.e., that medicine is normative precisely because it is a social practice, and as such is defined by the ends, or the social norms which that practice itself adopts or generates. And yet, I have just stated that medicine is essentially normative because its object, i.e., the functioning or healthy human being is an essentially normative being, and that therefore a proper account of the human being has to recognize the states of this being as better or worse in some respects that concern medicine. This somehow seems to be a confusion if not a contradiction in the argument, for if medicine is a normative practice because its object is *as such* a normative being then the normative character of medical knowledge would not derive from the social nature of this knowledge, but rather from some ground extrinsic to the practice itself. The *telos* or aim of the practice would somehow lie beyond the practice itself. The norms of medicine would have some transcendental ground in the nature of the person, or in the normative states of the person taken to be the object of medicine. The practical effects of such a view are clearly felt in medicine, where normative questions such as those concerning euthanasia, abortion, right to practice, and right to medical treatment are assigned to practices beyond medicine, for the establishment of norms, e.g. to, legal or judicial practice, or to theology, or to philosophy, or indeed are assigned to some determination in terms of what the prevailing moral attitudes and principles of the community happen to be. Does this mean that the ends of medicine, or what defines it as a practice, is not the social nature of medical knowledge, but something else? That the practice is not autonomously self-constituted? Yes and no. The concept of the person as a normative being is not one that is determined *apart* from the norms of medical knowledge; on the other hand, it is the very social character of medical knowledge that makes such knowledge both a constitutive *and* a derived factor of social knowledge in general. That is to say, medicine both contributes to the social definition of the person that is current in the society at large and also is affected by the concept or the varying concepts of the person that are socially established, or that are formed in part by other prevailing practices, e.g., by law (with respect to the juridical concept of the person), by the social or human sciences, by moral theory, or by theology. The general point is this: that what is *objectively* normative about the human being who is the object of medicine, is so by virtue of the self-constituting character of this objectivity itself; that is, people are what they are as social beings largely because of what they take themselves to be. Other animals are largely not what they are in this respect, but rather what they are by biological circumstance, or in the case of domesticated animals, what they are taken to be (and how they have been bred or used) by human beings. Thus, though there is clearly the biological basis of human beings that is shared with animals, it is the transformation of this biology by social and cultural life that produces the new human species and gives it its distinctive historical character. The *prima facie* evidence for this difference in medicine is that medical practice with respect to human patients is fundamentally and qualitatively different from veterinary medicine. The relation between physician and patient, or as Sigerist says, "between the medical corps and society", is a relation among persons, a social relation, or in Aristotle's original sense, an ethical relation. Henry Sigerist's statement of this relation is clear and striking, and worth quoting:

I have shocked medical audiences more than once by saying that medicine is not so much a natural as a social science. The goal of medicine is social. It is not only the cure of disease, the restoration of an organism. The goal is to keep man adjusted to his environment as a useful member of society or to readjust him as the case may be. In order to do this, medicine constantly applies methods of science but the ultimate goal is social nevertheless. In every medical action there are always two parties involved, the physician and the patient, or in a broader sense, the medical corps and society. Medicine is nothing else than the manifold relations between these two groups. The history of medicine, therefore, cannot limit itself to the history of science, institutions, and characters of medicine, but must include the history of the patient in society, that of the physician, and the history of relations between physician and patient. History thus becomes social history, and I hope to be able to show you that such an approach is promising and can contribute to a better understanding of social problems of medicine that we are facing today ([1], p. 26).

It is this social relation between persons which I would take to be the fundamental social presupposition of medical knowledge, both with respect to the acquisition of such knowledge, and to its use. But such a social relation involves, in a crucial way, the question of rights. It is in this sense that I have proposed that the epistemology of medicine essentially involves social theory, or that such an epistemology must be a social or historical epistemology.

II. KNOWLEDGE AND RIGHTS

There are two correlated sorts of rights that I am concerned with here: one is the right to practice medicine, and the other is the right to medical treatment or care. The second is related to, but is not identical with the right to health. In order to discuss these questions, I will digress first to a

more general question, namely, *Who has rights*? Let me begin obliquely
by posing the question in theological terms: God has no rights; more
specifically, God has no right to practice medicine and also no right to
medical treatment or to health. The reasons are simple: God has no rights
because there is no one against whom God could claim the right, no one
who could deprive God of anything He (or She) needs, wants, or desires.
God has no right to health, moreover, because God neither gets sick nor
suffers. God, as the theologians tell us, is impassible. After all, what kind
of a God would it be who could have needs, or require rights, or who
could get sick?

Now let me descend from heaven to earth, and put this less obliquely.
Rights are social, or to put it another way, rights are social artifacts. They
are created, sustained, violated, and lost by human beings in their relations to each other. Moreover, rights hold only among individuals who,
in this regard at least, stand to each other in a relation of reciprocity and
equality. For where x has a right with respect to y, then y has a reciprocal
duty or obligation to x. Rights therefore exist only in a network of social
relations, where they are correlated with duties or obligations. I cannot
be said to have a right which no one is bound to honor. Such a "right"
would be completely empty or vacuous – in effect, a pseudo-right. Or to
put it differently, it would be an abstract right, taken out of all connection
with the conditions under which it could be realized, or could become a
concrete right.

Further, if we take the right to practice medicine, or the right to
medical treatment as instances of rights, then this presupposes that the
practice of medicine or medical treatment is the sort of thing that
requires a right to sustain it, or to enable it, or to realize it. That is, such
a right constitutes a claim against someone else, and therefore the correlative duty on the part of someone else, in order for the right to be maintained or realized. Thus, to say that God has no rights is to say that God
has no peers against whom the right could be claimed. To say that one
has a right, therefore, is to say that there are peers against whom this
right could be claimed, and whose actions could realize it.

This general discussion then frames the question concerning rights
in medicine in a specific way: Against whom does the physician
claim the right to practice? And why does the practice of medicine
require that there be such a right? Correlatively, against whom does
the patient claim the right to medical services? And why does such
service require a right in order for it to be realized or given? This

is a different sort of question than the one that asks what are the *grounds* of such rights; for the position I am taking here already presupposes that all rights are grounded in social relations between persons, and that therefore, it is the agreement of these persons as to rights and duties that constitutes the ground for rights. This is what I mean when I say that rights are social artifacts.

To state my thesis briefly at the outset, I would argue that the physician's right to practice is a claim against society as a whole, and that its authorization as a right is based on the social need (or the socially *recognized* need) for the practice of medicine. But it is this same socially recognized need that is the basis for the patient's right to treatment. And yet, the patient's claim to this right is not simply a claim against society as a whole, but rather a claim against the medical profession in particular. Let me unravel this rather tangled knot. The ground for both of these rights is the socially recognized need for medical practice; but this need itself is grounded in the socially recognized need for health (however this comes to be defined from one society or culture to the next) and on the historical development of various modes of meeting this need, most specifically, upon the division of labor that leads to the practice of medicine, and finally to the professionalization of this practice, as it is embodied in the various historical forms that it takes. Broadly speaking, medicine is the distinctive social adaptation which our species has evolved, in its cultural history, for the maintenance of health and the management of disease. Thus, as I have argued, medicine in its curative and preventive aspects is a normative practice, whose norms are constituted both by the autonomous developments of medical knowledge concerning health and disease as normative facts, and by the social determinations of needs in these respects. But if it is the socially recognized needs that give rise to the requirement for medical practice as a way to meet these needs, then the society has a correlative duty to the physician, or to the medical profession to make available the conditions or means necessary to realize this right to practice. For without these conditions, the right remains abstract, and the practice cannot be carried out. These conditions include not only the material necessities for practice, e.g., the earning of a livelihood, access to the materials needed for the practice itself, but also the social conditions, i.e., the institutional forms of the practice as socially recognized and accepted forms, the recognition and authorization of the physician's professional decisions, the judicial recognition of the physician's rights, and his or her protection against undue interference with the practice. In short, the freedom of the physician to practice, and to be able to satisfy the needs of society is what is established by the right to practice; and this freedom is not simply some innate *or priori* quality that the physician has, but rather a socially established condition for the possibility of the practice itself.

But the right to practice, precisely in the sense that it is a right grounded on a social need, carries with it the responsibility of satisfying that need. Therefore the condition of the right itself is the physician's competence to practice in such a way as to satisfy that need. The authority of the physician that the right grants is an authority based on what the physician is able to do, or in the broad sense I have specified earlier, on the physician's knowledge. The abrogation of this right is therefore feasible only if the physician cannot responsibly exercise this knowledge, i.e., if the physician is incompetent. This suggests that the physician, as a condition for having the right to practice, has a duty with respect to some other right, which makes a claim upon him or her. Presumably, this is the right that society has to the services of medicine, which right in turn is socially established by the recognition that health and the treatment of disease are necessary aspects of social functioning. But it would then seem that the right to health, or to medical treatment, is a right that society holds against itself. In a sense this is true. It is an expression of the social interdependence that characterizes human life, that individuals can in effect make claims upon themselves, equally, in making social claims upon each other. Where such a claim is universal (or is held to be universal), that is, where each one holds this claim equally against all others, mutually, this, mutual interdependence, or mutual need generates what we may call a "human right", i.e., a right universally acknowledged as held by all the members of a society. But if the right to health, or to medical treatment is such a universal right, and is therefore a claim that any member of society may make upon all the others, it remains an empty or abstract right unless the conditions for meeting that claim can be concretely satisfied; and they cannot without the mediation of the practice of medicine. In this sense, medical practice is the mediating instrumentality or means by which a society realizes a right that each of its members hold with respect to all the others. In other words, the right to practice medicine is a fiduciary right, which is invested in the medical profession as society's representative instrumentality for realizing a socially recognized need. What is fiduciary about it is that the conditions for the entrusting of this function is that it be carried out in accordance with the needs of the society. But here, medical knowledge becomes not simply a condition for carrying forth a practice required by society, but becomes itself one of the central constitutive agencies in establishing what that requirement itself is. That is, medical knowledge itself comes to be crucial in the process of social recognition of the needs that the profession is set up to meet. Medicine thus not only does what society needs it to do, but comes to inform society what its needs are. In this sense, medicine is in part a self-constituting practice, defining its own ends, but against the larger social background of the need for health, and of the right to health. Medical knowledge thus comes to be not only the authorization for the right to practice, as the competence of one's practice is the correlative duty of the physician with respect to society's right to the services of medicine; beyond this, such knowledge also becomes the basis for the criterion of competence itself. In this respect, medicine has been historically given the right to authorize its own competence. But it has been granted this right in conditional ways, namely, on the condition that such self-authorization would in fact yield the benefits to society that it expects from medicine. In the old days, the doctor who failed to met these requirements was put to death. In modern times, certification and decertification become functions in which non-medical regulative agen-cies (whether Kings and Princes, in granting privileges to practice and withdrawing them, or State Boards of Certification as governmental bodies) have something to say.

In these ways, what the doctor knows is fundamentally bound up with what his or her professional rights are, and also with what claims against this knowledge the patient may make. Granted, this is a broad context within which to examine the specific relations of rights between physi-cians and patients, of the one to practice, and of the other to receive treatment or care. But if my premise, that rights are social artifacts and exist between persons in social relations, is correct, then we cannot go outside this relation to find the ground for such rights. What needs to be added is that these relations of rights and duties change historically, and that they are in the process of fundamental change in the present. But this is the subject of another paper. What is yet to be discussed here is the specific form this relation between persons, which is a rights-laden rela-tion, takes in the contexts of diagnosis. What, then is the relation between medical knowledge and rights as it is expressed in the diagnos-tic situation?

III. THE PHYSICIAN-PATIENT RELATION IN DIAGNOSIS

The claim I propose to support in this final section of the paper is that the science of diagnosis cannot be properly understood, either epistemologically or methodologically, apart from the normative contexts of medical practice; specifically, that a theory of clinical judgment is essentially incomplete if it leaves out of account the relations between physician and patient as these involve the respective rights and duties of each of the parties.

Insofar as the diagnostic process is a case of scientific judgment, it concerns an analysis of how the symptoms, signs and complaints that a patient presents are to be interpreted by the diagnostician in order to reach a conclusion that identifies correctly the pathology that is the cause of the patient's condition. The aim of diagnosis, however, is not simply to reach such a conclusion. Rather, the conclusion is for the sake of a therapeutic judgment to be made on the basis of the diagnosis. Diagnosis is therefore not a closed procedure, but rather an open-ended one, as a stage in the process of clinical judgment. Review and revision of a diagnostic conclusion may follow upon the results of some therapeutic intervention; or indeed, the conclusion reached as the result of a diagnosis may be one of "watchful waiting" to see if a finer determination can be made as the result of further developments, or indeed to see if a prognosis is correct, where the diagnosis itself leaves room for doubt or where the indicated therapy is so radical or risky that further assurance about the diagnosis may reasonably be sought. The scientific character of the diagnostic judgment involves the use of trained observation, physical examination, laboratory tests, and skillful medical history taking to establish the data for a judgment; but it involves equally the trained use of reasoning, in the framing of hypotheses, the deduction of the consequences of such hypotheses, and the testing of such deductive consequences against the empirical facts. In these respects, the making of a diagnostic judgments has all the hallmarks of scientific judgment in general. What distinguishes the diagnostic judgment from judgments in the natural sciences, however, is not the additional factor of what is sometimes referred to as clinical intuition, or an ability to make a gestalt-like or a *Verstehen* sort of judgment, for I would argue that this feature holds as much in the natural sciences as it does in the social sciences, or in clinical judgments in medicine. However one may analyze this recalcitrant feature of clinical or scientific judgment, it is not peculiar to medicine. What is peculiar to the diagnostic judgment beyond these general scientific features is that it is essentially a casuistical judgment; that is, it is an interpretation of general principles, in the light of clinical experience, in application to a particular or individual case. In this respect it is analogous to judicial judgments in law, to aesthetic judgment in the arts, or to practical judgment in moral and personal contexts. The question is: What shall I decide about this particular case, about this individual? The facile solution, which seems to follow from the diagnostic textbook, is to establish that this individual case is one of a given type or class or cases; and once this is established, what holds for the class holds for the members of that class. Even common sensically, it would seem that if it can be established that something is a case of diabetes, or of carcinoma of the lung, then the diagnostic conclusion is complete, and what follows is prognosis and therapy. This, as one of my students replied, will get you as far as third year in medical school. (He was a fourth year student.) My argument is, first, that this is not the conclusion of the diagnostic process; and second, that it gives an abstract reconstruction of the process in a way which leaves out of account an essential component, if not an overriding characteristic of the process as a whole. The feature that makes diagnosis a distinctive form of judgment is that its object is a specific individual, with a history, a present and a future, each of which bear upon the diagnosis itself; and not simply with respect to the "bedside manner" of the physician, but with respect to the process of scientific judgment itself. One aspect of this, of course, is the unique individuality of the patient. Disease agents may be generalizable. That is, all pneumococci are pneumococci. But susceptibility is highly differential. So, too is the particular complex of associated pathologies which may be coexistent with a given identifiable one, so that the diagnostic sharp focus is subject to smearing out by such complications. And these are highly differential among individuals. A second aspect of this individuality of the diagnostic subject is the role which the patient's own complaints, descriptions, answers to questions, and indeed the patient's observable behavior, demeanor, and even clinically measurable sensitivity to being examined plays in the diagnostic process. Here, there is a spectrum of vulnerability to examination: blood pressure may be more responsive to the stress of the diagnostic situation than bone tissue is; and in general, psychological response is likely to be more volatile than somatic response (though a case could be made either way). A third aspect of this individuality is on the other end of the stethoscope, so to speak: the diagnostician is also an individual, acting in a social relation (or as the microsociologists like to say, a "face-to-face interaction") with another individual. The individuality of the physician here includes not only the obvious questions of temperament, prejudice, mood, state of health or of attention, but the more explicit cognitive differentials of training, specialty, background knowledge, and even cognitive style – whether emphatically empiricist and inductivist, or conjectural, or intuitive, or painfully deductive.

In short, the relation between physician and patient in the diagnostic process is, even in the most explicit "scientific" reaches of the process, one which inevitably involves personal relations between individuals. Not so, one may say, for the pathologist or the radiologist or the laboratory technician who is dealing with relativity anonymous samples of blood, tissue or fluid, or with information generated by one or another diagnostic technology, in the most impersonal way. True enough. But the end point of the diagnostic stage has still to be reached with respect to a given and unique individual, whatever depersonalization there may be, for good scientific reasons, in the auxiliary components of the diagnosis.

But if, indeed, the relation between diagnostician and diagnostic subject is a relation among persons, specifically between physician and patient, then there is still more to be taken into account. Tacitly or explicity, the relation between the two is one which involves just that relation of rights and duties which I discussed earlier. Now I could mean the general rights and duties which concern the social relations between medical practice and society, as they are instantiated in *any* medical relation. But here, I want to go beyond this, to the specific rights and duties involved in the diagnostic process, *as a process of clinical investigation.* Here the right of the physician to perform the diagnosis becomes a concrete right only if the patient (or the patient's proxy, where paternalist contexts call for one) has a corresponding duty to the diagnostician. So, for example, the physician's right to make certain tests, or to perform a physical examination of a certain sort entails the patient's duty to have these tests or this examination performed. Is this a duty owed to the physician? By our previous analysis, this duty is owed *if* the patient has a right to medical care; for this right is based on a need, the satisfaction of which requires competent medical action. Thus the physician's right, here too, remains abstract if the cooperative obligation of the patient to observe this right is withheld, and then the patient cannot exercise his or her own right to medical treatment unless the physician is granted the correlative right to practice in the concrete case.

There is nothing abstract about this in diagnostic terms. For if the patient does not dutifully cooperate in the examination, then the diagnostic judgment itself is put in jeopardy to the extent that what the patient does by way of contributing to the required information affects the basis for the judgment. Similarly, the patient's right remains abstract unless the competence of the diagnostician is adequate to the task. What the patient has a right to is not simply a diagnosis, but a competent one, and at the limit, a correct one. The normative expectations are here crucial to a characterization of the diagnostic process itself. This may be obvious, tacit, and well-understood – namely, that the physician is expected to do the best he or she can on the basis of a satisfactorily competent level of medical knowledge and skill – but this is not simply something the physician demands of himself or herself, or which is demanded by the standards of the profession. Rather, it is demanded by the rights which patients have, in the social relations with physicians, however these rights may come to be institutionalized in intra-professional criteria of competence.

The subject, in any case, needs further analysis and interpretation in the concrete contexts of diagnostic procedure. All that I have attempted here is a sketch of how such clinical judgment is implicated in the network of personal relations between physician and patient, not simply in a social-theoretical sense, but in explicitly epistemological terms. In this sense, the ethics of diagnosis is at one and at the same time an epistemological and a value-theoretical question, and one in which these two components cannot be separated without injustice to the very nature of the diagnostic process.

The City University of New York
Baruch College and the Graduate Center

BIBLIOGRAPHY

1. Sigerist, H.E.: 1960, *Henry E. Sigerist On the History of Medicine*, In F. Marti-Ibanez (ed.), Md Publications, Inc., New York.

HANS-MARTIN SASS

DIAGNOSING THE ELEVEN MONTH PREGNANCY: SOME ASPECTS OF MORAL AND CULTURAL FOUNDATIONS IN MEDICAL JUDGMENT

I. DIAGNOSING BRAIN DEATH AND ELEVEN MONTH PREGNANCY

François Rabelais (1483–1553), Professor of Medicine at the prestigious Université de Montpellier in 1537, had a literary offspring named Gargantua. Gargantua was born in Lyon, after gestationing eleven months, in the fourth chapter of Dr. Rabelais's "Vie inestimable du grand Gargantua, père du Pantagruel" ([16],[21]). The eleven month pregnancy provided Gargantua with a long and adventurous life in four subsequent volumes; and he is still very much alive in the history of ideas.

Why did Dr. Rabelais, as so many other physicians of his time, diagnose an eleven month pregnancy? He might have had moral or cultural reasons for declaring an offspring born as late as eleven months after the last sexual intercourse of a couple to be a legitimate child. During the early sixteenth century intestate succession was a complex legal, political and personal problem in cases when the feudal testator died without an heir. In such cases the estate would pass to a male relative of the testator, not to his wife. Diagnosing an eleven month pregnancy can be understood as an attempt to solve, or at least, to minimize the risks to the widow and her economic and cultural well-being by giving her eight weeks additional to the "normal" nine months after the death or departure of her husband to give birth to an heir. Carrenza, in 1630 ([4], p. 582), stated that the eleven month period is the longest scientifically acceptable time frame for pregnancy. He quotes the Hippocratic Corpus and also Aristotle's "Peri zoon geneseos", discussing seven, eight, and ten month pregnancies together with the biological law of animation, i.e., the soul entering the fetal body just at the end of the fortieth day of pregnancy or, should the foetus be female, after the eightieth day, making it a morally recognizable human being only thereafter and eliminating the moral problems of abortion prior to animation, a period during which abortion was of low risk to the mother (Aristotle, "Historia Animalium", VIII, 3 5838).

J. L. Peset and D. Gracia (eds.), The Ethics of Diagnosis, 153–162.
© 1992 *by Kluwer Academic Publishers. Printed in the Netherlands.*

Indeed, the biological laws of a relatively wide scale in the human pregnancy period and the animation law, when used in medical diagnosis, serve as a moral and cultural risk reduction instrument for the public moral, legal and cultural debate, and also as an instrument to benefit the patient far beyond the more narrow concept of medical treatment of diseases.

What is called a biological law exoterically by the medical profession is, in fact, an esoteric value judgment established within the medical profession, hidden from the public and intended to serve the well-being of the public and the patient in the best possible way, including the protection of the publicly established value system, honoring correct and inviolable rules of intestate succession, not killing human persons, etc. The *Medicorum Communis,* wedding expertise and ethics into professional judgment and exercising professional paternalism over society and its established moral and cultural goods, abuses "science" as a professional insight into the laws of nature by defining "laws of nature" within a not purely scientific, but rather a diagnostic and therapeutic professional activity. Such a concept of professional responsibility sheds light on traditional priorities within the medical profession regarding the duties or commitments towards "scientific truth", "pure science", "the patient", and "the society".

Wieland describes the historical process from Sydenham towards the contemporary concept of diagnosis ([29]), pp. 106, 119) as a way away from ontological schemes of classification of "species morbosae" towards a more flexible interpretation. It should be added, however, that for both systems: the diagnostic tool of classified substantialized concepts of disease as well as modern forms of data interpreting simple or differential diagnosis, have their moral and cultural implications. As the Gargantua case demonstrates, there are ways around the substance concept of disease and "natural normality", in fact, by using this concept for pushing professional moral interventions esoterically. In our times what Wieland calls "flexible diagnosis of data collected" has led to the language of "therapeutic abortions" (*Beal v. Doe*, 423 US 438, 1977), or "social indication" for diagnosing the need for abortion, such as the threat of the mother to commit suicide in case an abortion is refused (Federal Republic of Germany, Penal Code § 218) or diagnosing the moral value of the fetus as compared to the "right to privacy" of the mother, which would diagnose unwanted pregnancy as an infringement on the well-being of the mother and therefore exercise adequate therapy, i.e., abortion (Roe vs. Wade, 410 U.S. 113 (1973), prenatally or diagnosing "handicaps" which would or would not allow or invite abortion. Also, what we call "brain death" more appropriately would use a more precise, less evaluative language, calling the diagnosis "cessation of brain activity", of all or major parts of the "brain organ", with highest possible likelihood of irreversibility. Calling the irreversible non-functioning of the brain tissue death involves no less nontechnical ingredients in diagnosis than does diagnosing eleven month pregnancies.

There are less sensational cultural and moral implications in medical diagnosis such as in diagnosing syndromes, hypertonia, vegetative dystonia, macho behavior as abnormal or feminism as abnormal, masturbation as a disease. Whining and weeping was a well accepted pattern for educated man in the European Romantic culture of the early nineteenth century, it is not for the educated corporate man in our contemporary culture. As far as physicians do belong to their contemporary culture, changing cultural and moral priorities will, more or less, influence the formation of diagnosis. The influence will be higher if physicians do no continue to have or establish a separate professional culture, apart and separate from the mainstream of public culture. The influence will be less if physicians do adhere to a specifically medical tradition and value system. But, then again, the foundation of the physician's distance towards the public culture will be a cultural one, and not a scientific one, even through the role of science and its internal ethos definitely are part of the culture [17].

II. PERILS OF GENERALIZATION OF SCIENTIFIC DATA

Post Enlightenment times, having destroyed the domains, kingdoms and townships of metaphysically or religiously established absolute value, feel tempted to replace these sacred cows by new ones, the edifices of scientific generalization. These generalizations and ideologies serve the purpose of an essay orientation and understanding of "reality" and the predispositioning of moral behavior. Science, in the form of biology, has served as such a generalization model for ideologies, for example, Darwin's "struggle for life" as a blueprint for Social Darwinism, and Darwin's remarks on "mutual aid" as a blueprint for Anarchism à la Kropotkin [14]. Lysenko's theory on the inheritance of learned behavior backed a totalitarian system of education [15], while Skinner's research on the behavior of rats has been transferred unrestrictedly to human sociology [23]. Other examples for ideological generalizations are based on economy, psychology, or sociology. These generalizations, apart from their logical and intellectual shortcomings and fallacies, represent *moral malpractice*, as ideologies just replace God or other metaphysical supreme powers for data or schemes of one of the sciences [19]. The system of reference, basing moral and cultural values on something substantial, eternal has not been changed in meta-scientific ideologies, replacing metaphysical dogmas. It would have been changed, if values would be accepted on their own merits, for example, in con-tractualism, utilitarianism or based on the authority of cultural tradition.

Culture, not nature, is the basic foundation of values. Those values include the moral and cultural values that play an important role in seeing the social and cultural framework of physician-patient relationships, in providing essential structural ingredients in diagnosis, in setting the medical profession in some way apart from the public value system and making them, on the other hand, a part if it. The original Hippocratic Oath is a good paradigm for the complex cultural roots of medical judgment. Today professional codes of conduct, self regulation by medical societies, internal guidelines, and hospital policies do shape the struc-ture and the preconditions of the diagnosis. Ethics Committees supplement the interpretation of data in cases in which the individual physician feels overbur-dened or is expected to be overburdened by the complexity of the diagnosis. I do not claim, however, that diagnosis is nothing else than the application of human-ist values in a moral or cultural risk assessment situation. No good diagnosis can be done without the auxiliary help of various sciences such as hemotology, radi-ology, pharmacology, sociology, and psychology. But all the data collected by these sciences do not yet make a diagnosis. The auxiliary use of applied sciences and applied technologies does not make diagnosis a pure scientific or technical judgment [20]. Engelhardt describes impressively how the language of medicine, that is, the software of diagnosis, has changed over the centuries and how evalua-tive, descriptive, and explanatory elements have determined the nature of diagno-sis, which shapes the social reality of the physician-patient relationship ([8], pp. 164–184). The complexity of judgment in diagnosis makes it difficult to under-stand the nature of medical risk analysis and risk assessment but it makes it easy to grasp the point that the reality of diagnosis – its responsibility structure, its intertwining of analytical data, cultural demands, traditional role models, esoteric value tradition, and the overall ethos of healing and comforting, is superior to the all too simple generalizations and ontologizations of scientific data from different sources [20].

III. FROM SPECIES MORBIDAE TOWARDS COMPUTER DIAGNOSIS

Douglas and Wildavsky have argued that risks are altogether "collective con-structs" [6] without any basis in reality outside the world of collective constructs.

structs. This might be the case for some highly fashionable and, from time to time, changing arbitrary selections and preferences in social or cultural risk awareness. It is not so, however, with regard to health risk for mortal human beings. But certain forms of health risks enjoy a high degree of social and personal awareness, elevating those risks into the canon of commonly accepted and described forms of disease, as in the case of masturbation.

T.H. Huxley, in the Romanes Lecture (1893), sharply criticized the concept of nature and evolution as a tool for moral and cultural enrichment and orientation. He confronted the state of nature and biology with the state of humans and culture; the latter does not rest on the laws of nature, but surpasses them and contradicts them: "Cosmic evolution may teach us how the good and the evil tendencies of man have come about; but in itself, it is incompetent to furnish any better reason why what we call good is preferable to what we call evil" ([11], p. 80). The ethical progress, then, depends "not on imitating the cosmic process, still less in running away from it, but in combating it. Human culture and cultivated humans need succeeded in building up an arfiticial world within the cosmos", i.e., the cultural world ([11], p. 82f). with regard to the foundations and goals of diagnosis, cultural and moral implications may be seen to play and essential part in – describing what is good or evil concerning "health". Joseph Fuchs' analysis of the Christian *lex naturale* tradition came to a similar position when it argued that "divine law" can only mean a "moral code which God has implanted as law into the heart of man" ([9], pp. 369). Human authority in this world, he argued, cannot just rest on a few global laws, but needs detailed insights (*Einsichten*), which cannot simply be understood as "application of 'divine laws'" ([9], p. 382). Therefore, he concluded, the "application" concept of divine laws in concrete cases of everyday life is of very limited practicability and relevance. This Christian concept of diagnosis does not argue on the grounds of applying general divine laws in clinical judgment, but rather calls for understanding human dignity and responsibility is a concrete and actual situation of a particular case. On the basis of such "ethics of moral action within this world" (*Moral innerweltlichen Hadelns*), there cannot be any prefabricated position of Christian morality with regard to controversial issues such as premarital sex, masturbation, brain death definition, or the moral states of early stages of human gestation ([9], pp. 317–373). Rather, the moral mastery of these issues has to come through concrete responsibility of existing humans within the world. Diagnosis is always part of physician-patient communication. As patients are not primarily interested in classifications or microscopes, but in symptoms and in treatment and in its results and side-effects, this side of the diagnosis shows even more convincingly that, in the words of Shelp, "all facts, including illness and suffering, are constituted for us in terms of a rich set of cultural and social expectations" ([22], p. 255).

The history of modern diagnosis demonstrates a development from the onto-logical classification scheme of the *species morbidae* [25] and forms of causal argumentation in diagnosis to addressing only the "components, which constitute a specific disease" ([29], p. 129). The professional ethos of the physician, devo-tion to comforting, healing, and supporting patients – and not to establishing or inventing eternal ontological, biological, or medical laws and classifications – was instrumental in establishing a pragmatic approach in diagnosis not resting on the table of *species morbida* or theological generalizations.

Koch defined diagnosis as "the sum of knowledge obtained and instrumental in the physician's actions and conduct". This knowledge may include information other than those provided by the classical diagnosis and may result in "diag-nosoids" [12]. Both for reasons of limited efficacy of the classification diagnosis and for moral reasons, the teaching and applying of the *species morbida* concept becomes replaced by the analysis and diagnosis of a "combination of factors" ([23], p. 132).

Diagnosis becomes an expertise in "analyzing factors" (Faktoren-analyse) ([27]; [29], p. 134). Braun puts the ontological aspects of clinical diagnosis totally aside and develops a diagnostic "decision theory"; the practice of diagno-sis uses "terms of the theory of the profession" (Berufstheoretische Begriffe). The most basic decision is to diagnose either an "avertable or preventable dan-gerous process" (abwendbarer gefaehrlicher Verlauf) or to decide about "cau-tiously keeping the decision open" (abwartednes Offenlassen) [3]. In both cases classical descriptions of disease may be consulted, but they do not predetermine the course of medical action (intervention or non-intervention). Wieland ([29], p. 171) underlines the pragmatic dimension of diagnosis which he calls a "tempo-rary singular statement" (zeitgebundenen Singulaeraussage) within which the classical theory of substantial *species morbida* serves as one among other pat-terns of reference in clinical judgment.

The development of European concepts of diagnosis from Sydenham's classification to Braun's and Wieland's pragmatic approach was foreshadowed by John Brown's (1735–1788) critique of Sydenham and his diagnostic theory of stimuli and sedativa, of a tension between stimulus and excitement [3]. Overexcitement causes "sthenic diseases", underexcitement causes "asthenic dis-eases" ([24], p. 61). Brownism is an excellent example of a revolutionary exchange of paradigm in scientific orientation ([13], [24]); external as well as internal factors in medicine were responsible for the lack of success of Brown's "nonclassification concept of diagnosis" (ie., identifying various stages or intensi-ties of excitability and insitability).

But the fact that the *species morbidae* theory finally is undestood as inadequate to serve as the sole foundation for diagnosis demonstrates that clinical judgment is a preeminently practical endeavor, which goes beyond the boundaries of formal classifications and involves decision-making without the comfortable framework of easily available natural laws and their scientific mastery. Inasmuch as this *primacy of praxis* is concerned, there is a variety of roots for medical diagnosis, one among others is moral concern and moral judgment.

IV. PROFESSIONALLY DIAGNOSING RISKS AND THE RIGHT TO CHOOSE

Wartofsky understands the medical judgment as a form of social practice and the values involved in such a social practice, in turn, are generated by socially regenerating cultural and moral needs and value priorities [28]. He concludes that diagnosis is more of a social judgment than a scientific judgment, not only with regard to the use of medical knowledge but also with regard to its roots and acquisition. If he is correct, and I think he is much closer to the reality of the everyday practice of diagnosis than positions identifying diagnosis exclusively with natural science, then the epistemologically and morally most pressing issue is the physician-patient interaction. Is it a consumer-provider relationship, not really different from other relationships in a society rich in a plurality of valuables and values in which the educated consumer may choose according to his needs, limited only by limits to financial resources or imaginative forces? Or, as health is something fundamental and preconditional to other goods, goals, enjoyments and achievements in life, is there still a role for professional paternalism in shaping the language of diagnosis? Whatever side we taken, we would make a value judgment, not a scientific conclusion.

Informed consent has been introduced and used as an instrument to protect the value of privacy of the patient as well as to solve the conflict between paternalism and uneducated consumerism. However, as the quality of consent is rather difficult to evaluate in situations of dependency, pain risk of opportunities or life, informed consent as a normative instrument seems to be only of auxiliary and pragmatic value. Also, it puts the evaluative issues into the physician-patient relationship prior to therapy but subsequent to diagnosis, which is understood as descriptive rather than evaluative. But, as we have discussed earlier, value judgments are already crucially involved in the process of diagnosis. They could be reduced by stripping diagnosis as much as possible of all the ingredients other than computer data and data relationships. But then, what would be the quality of information into the informed decision making communication with the patient?

Gethmann recommends the physician to have a "fictive dialogue" with the patent, neither a real one seeking a compromise between possible conflicting interests, nor professional application of general expertise to a singular case without dialogue. Such a dialogue would potentially need to include the entire history and personal lifestyle of the patient ([10], p. 158). Gethman argues that factual acceptance might lack original rational consent while, on the other hand, the issue of therapy is not one which can be described simply in scientific terms. In introducing the *fictive discourse*, he offers a case-oriented golden rule, recommending normatively what can be described empirically as the practice of the good doctor anyway. However, the evaluative problem does not arise as late or in the transition from diagnosis to therapy. It is already an intrinsic factor within diagnosis.

Wartofsky describes the role of the physician as "society's representative instrumentality for realizing a socially accepted need" ([28], p. 146). His or her "right" to practice medicine therefore is a "fiduciary right". Using the rights language, he correlates the patient's duty to cooperate in the examination and gathering of data with the physician's duty to perform at a satisfactorily competent level of professionalism. This would best be undertaken under a concept of "gestalt" like or "verstehen" sort of judgment. It would necessarily include "personal relations between individuals".

Wartofsky seems to favor a dialectical dialogue, structured à la Habermas and Apel, between physician and patient within the diagnostic process. On the other hand, he admits that the rules for such a communication, protecting patient's rights and physician's fiduciary rights, might be "institutionalized in intra-professional criteria of competence" ([28], p. 151).

Waging professional expertise versus consumer lifestyle related needs in the process of diagnosis, it would be helpful to differentiate between the "scientific data" and the "evaluative" design of collecting data, evaluating data and formulating the diagnosis. Also, it would be mandatory in the data gathering process to not only collect *natural data* (x-ray, blood test, etc.) but also *cultural data* (risk acceptance, value priorities, etc.) within the process of diagnosis [17]. Thus, not striving for eliminating nontechnical cultural data out of the diagnostic process, but rather *accentuating* them and giving them an essential role already in diagnosis, equal to the collection and assessment of technical data, would be the final goal of a good diagnosis. Elsewhere, I have proposed the development of ethics modules or medical ethical metalanguages for interactive expert systems [19] in medicine. This would allow for the integration of axioscopic aspects and of medical-ethical diagnosis and prognosis into medical-technical diagnosis and prognosis, for the *integration of ethics and expertise* in computer age medicine. The *Bochum Questionnaire for Medical Ethical Practice* [20], a checklist for clinical single case treatment differentiates between medical-technical diagnosis and medical-ethical diagnosis and integrates the results of both into prognosis and therapy decisions; major issues in medical-ethical diagnosis are quality of life, patient autonomy, and physicians responsibility.

Integrating the *two sets of information* into the diagnostic process would (1) help eliminate or decrease professional paternalism, (2) avoid the narrowing down of diagnosis to gathering of technical data, (3) prepare a prudent physician-patient communication including lifestyle related value priorities, and (4) lead into educated (not just informed) judgment, decision-making, or consent. We have a German saying, "Wer heilt, hat recht" – "who heals is right." As the answer to the question "what is healing" depends on a plurality of options to define well-being, well-feeling, not feeling well personally, culturally, medically "right" in this context is the "right or wrong" in neither the scientific nor in the legal sense. It is the "adequance" within an individual case. That was the reason seventeenth century physicians diagnosed eleven month pregnancies, taking natural (biological) *and* cultural (meta-biological) data into account. The value of their diagnosis could be right or wrong, but only right or wrong in a supra-scientific, supra-legal even supra-moral aspect. It would be right or wrong according to professional competence, wedding technical expertise and moral ethos.

Institut für Philosophie
Ruhr-Universität Bochun
Bochum, Germany
Georgetown University
Washington, D.C. U.S.A

BIBLIOGRAPHY

1. Albertus Magnus: 1921, *De animalibus*, H. Stadler, (ed.) Aschendorff, Münster.
2. Braun, R.N.: 1970, *Lehrbuch der ärztlichen Allgemeinpraxis*, Urban and Schwarzenberg, München.
3. Brown, J.: 1806, *Anfangsgründe der Medizen*, A Roeschlaub (ed.), Frankfurt, Germany.
4. Carranza, A.A.: 1630, *De partu naturali et legitimo*, Geneva, Switzerland.
5. Darwin, C.: 1859, *The Origin of Species by Means of Natural Selection*, 2nd. ed., John Murray, London, U.K.
6. Douglas, M., and Wildavsky, A.: 1982, *Risk and Culture*, University of California Press, Los Angeles, California.
7. Engelhardt, H.T.: 1974, 'The Disease of Masturbation. Values and the Concept of Disease', *Bulletin of the History of Medicine* **48**, 234–248.
8. Engelhardt, H.T.: 1986, *The Foundations of Bioethics*, Oxford University Press, Oxford.

162   HANS-MARTIN SASS

9. Fuchs, J.: 1984, 'Das Gottesbild Und die Moral innerweltlichen Handelns', *Stimmen der Zeit* **202**, Jg. 109 363–382.
10. Gethmann, C.F.: 1986, "Abzeptanz and Akzertabilitä von Risken," in M. Anlauf and K.D. bock (eds.), *Milde Hypertonie und leichte Fefstoffwechels* törengen, Steinkopf, Darmstadt, pp. 149–162.
11. Huxley, T.H.: 1947, (1893) 'Evolution and Ethics', *Evolution and Ethics 1893–1943*, The Pilot Press, London, U.K., pp. 61–84.
12. Koch, R.: 1920, *Die ärztiche Diagnose*, 2nd ed., J.F. Bergmann, Wiesbaden, Germany.
13. Kuhn, T.H.S.: 1973, *Die Struktur der wissenschaftlichen Revolution*, Suhrkamp, Frankfurt, Germany.
14. Kropotkin, P.A.: 1955,(1902), *Mutual Aid*, Extending Horizons Books, Boston, Massachusetts.
15. Medvedev, Z.A.: 1969, *The Rise and Fall of J.D. Lysenko* New York.
16. Rabelais, F.: 1970 (1534), *Vie inestimable du grand Gargantua, père du Panatguel*, H. Heintze and E. Heintze (eds.), German ed., Diederich, Leipzig.
17. Sass, H.M.: 1983, 'Standards in Technology and Human Values', *Wandlung von Verantwortungen und Werten in unser Zeit*, Saur Muchen, pp. 62–75, pp. 221–236.
18. Sass, H.M.: 1985, 'Zukünftige Entwicklungsmoeglichkeiten der extrakorporalen Fertilisation, ethische Aspekte', *Archives of Gynecology* **238**, 81–105.
19. Sass, H.M.: 1989, *Zur ethischen Bewertung von Expertensystemen in der Medizin*, Zentrum für Medizinische Ethik, Bochum (Medizinethische Materialien Nr. 44).
20. Sass, H.M. and H. Viefhues: 1989, 'Bochumer Arbeitsbogen für die Medizinethische Praxis', *Medizin und Ethik*, H.M. Sass (ed.), Reclam, Stuttgart, 371–375.
21. Screech, M.A.: 1969, 'Eleven Month Pregnancies, '*Etudes Rabelaisiennes*, **7**, 94–106.
22. Shelp, E.E.: 1984, 'The Experience of Illness', *The Journal of Medicine and Philosophy* **9**, 253–256.
23. Skinner, B.F.: 1971, *Beyond Freedom and Dignity*, Knopf, New York.
24. Schwanitz, H.J.: 1983, *Homeopathie and Brownianismus 1795–1844*, Fischer, Stuttgart.
25. Sydenham, Th.: 1846, *Opera Omnia*, G.A. Greenhill (ed.), Societas Sydenham, London.
26. Tettamanzi, D.: 1983, 'Verita ed Ethos', *L'Osservatore Romano*, Sept, **28**, 1983.
27. Ueberla, K.: 1971, *Faktorenanalyse. Eine systematische Einführung*, Winter, Heidelberg.
28. Wartofsky, M.A.: 1992, The Social Presuppositions of Medical Knowledge', in this volume, 131–151.
29. Wieland, W.: 1975, *Diagnose, Überlegungen zur Medizintheorie*, de Gruyter, Berlin, Germany.
30. Zubiri, X.: 1963, *Sobre la Esencia*, Soc. de Estudios, Madrid.

MARY ANN GARDELL CUTTER

VALUE PRESUPPOSITIONS OF DIAGNOSIS: A CASE STUDY IN DIAGNOSING CERVICAL CANCER

Through diagnosis a clinician approaches disorders by applying an explanatory account that allows the patient to be cast in a clinical category and therapy role. Because diagnosis usually involves understanding and undertaking clinical problems through pathoanatomical and pathophysiological frameworks, clinical diagnosis is theory-ladened. Diagnosis also involves judgments about the worthiness of particular conditions for special attention. These judgments involve evaluative considerations, and turn in part on accepted norms of scientific investigation and therapeutic success. As a result, diagnosis is pursued not solely for its own sake, but also for the sake of satisfying certain evaluative frameworks. In other words, clinicians seek to know well, as opposed to simply knowing truly the character of clinical problems. In short, diagnosis is contextual and intervention-oriented.

This essay explores the evaluative character of diagnosis, focussing on the non-moral and moral values involved in diagnosis and taking as its leading example issues involved in the diagnosis of cervical cancer using the Papanicolaou (Pap) smear. It argues that clinical diagnosis must be understood and undertaken within particular socio-cultural frameworks. This is to underscore the obvious, for medicine is an applied, not a pure, science, and hence involves a complex evaluative dimension.

In investigating the evaluative presuppositions of clinical diagnosis, one is introduced to major ambiguities in the terms 'diagnosis' and "evaluation". The term "diagnosis" (Greek, *diagnōsis*, discrimination) refers generally to a judgment that identifies particular clinical problems. Diagnostics endeavor to answer correctly the query "Why does a patient have *this* particular clinical manifestation as opposed to another?". Here "correctly" is understood as the use of good reason and prudence in the choice of one diagnosis over another. This use of diagnosis contrasts with many of the ways in which clinicians often speak of clinical diagnosis as if it were a form of intuition [7] rather than the outcome of a discursive rational process [13]. On this account, intuition may rather be understood as a capacity developed over many years of experience in diagnosing patient complaints. This capacity involves, as this essay

163

J. L. Peset and D. Gracia (eds.), The Ethics of Diagnosis, 163–170.
© 1992 *by Kluwer Academic Publishers. Printed in the Netherlands.*

shows, rational as well as evaluative considerations in distinquishing clinical problems.

Generally put, "to evaluate" means to assign significance. Evaluation, or the process of assigning a value, includes a wide variety of expressions of significance. Among these are non-moral (e.g., aesthetic, instrumental) and moral (i.e., normative) values. Non-moral values may be understood in terms of moral values. Moral values are those appealed to in an ethics. The term "ethics" trades on major ambiguities. Generally put, ethics as a philosophical discipline is devoted to the study of moral values, virtues, and the principles of action that bind individuals and provide common grounds for blame and praise and for delineating the character of the good life. Ethics has been understood in numerous senses, as the rules of conduct as determined by the customs of a people, by the etiquette of a profession, by the rules of law, and by the teachings of a religious doctrine. All of these contrast in different ways with the result of rational analysis, including both plausible and conclusive accounts of proper moral conduct [5], p. 54). In short, ethics has come to be associated with particular moral views of a culture, profession, legal body, or with the community of rational agents who determines what should be generally endorsed or normatively correct. It is this last sense of ethics that has been the focus of contemporary philosophical thought.

In addition to normative evaluation, a value may be placed on an object to express that the object is aesthetically pleasing or worthy of attention. An object is singled out in this way so as to bring it into focus for the perceiver. When an object of perception is singled out because it is seen as being useful for achieving a certain goal, consequence, or end state, instrumental evaluation takes place. Aesthetic and instrumental values vary among perceivers depending on particular desires, wishes, and goals. Herein lies the ambiguity: who decides what is aesthetically pleasing, what is beneficial, and what is a burden? How does one calculate beauty, benefit, and burden? This ambiguity leads to conflict in our practical lives. Alternatively, this ambiguity may be celebrated in the sense that rational decision-makers are free to hold different views of the good life. In order to celebrate, however, one needs to appreciate the tie between non-moral and moral evaluation. This tie is found at the interface between or among the game of art and/or the game of cost/benefit analysis with the game of choosing, the game of responsibility, and the game of praising and blaming.

In short, the language of evaluation and diagnosis is ambiguous. As is next discussed, this ambiguity influences the ways in which disease classifications are understood and undertaken. This is especially the case in the enterprise of classifying cancers.

DIAGNOSING CERVICAL CANCER

Cervical cancer is the third most frequent malignancy in women. Only carcinomas of the breast and of the endometrium are more common. It is estimated that about 2% of all women over age 40 will develop cervical cancer. About ninety percent of these neoplasms are squamous-cell carcinomas, and the remaining are adenocarcinomas and mixed adenosquamous carcinomas ([2], p. 243). The risk of this malignancy is increased in lower socio-economic groups, in those who have early marriages, in those who begin sexual activity or have children early, in those with many sexual partners, and in prostitutes [11]. Today, about 8000 women in the United States die each year of the disease. The hope is that given early detection and treatment, deaths due to cervical cancer can be significantly reduced.

The progression from the early stages, in which the cancer cells are localized to the outer layer of cells lining the uterine cervix, to the late stage, in which the cancer cells have invaded the underlying, muscular layers of the uterus, takes up to thirty-five years ([9], p. 204). Carcinoma of the cervix may be asymptotomatic in its early stages but may be detected by a routine Papanicolaou smear.[1] The Pap smear detects the presence of atypical cells found in the cervix and surrounding regions associated with clinical problems by a special staining technique. The properly collected Pap smear can accurately lead to the diagnosis of carcinoma of the cervix in about 98% of cases at best ([2], p. 122).

Clinicians' use of the Pap smear to diagnose cervical cancer reflects the ways in which clinicians agree on 1) how to acquire evidence relevant to the problem, and 2) how to reason with the evidence in order to reach a rationally defensible conclusion that will resolve some question or quandary. This realist or objective observer approach to explaining clinical problems presumes the possibility of a universal diagnosis for particular clinical problems. One is reminded here of the ways in which pathologists presumably describe cervical cancer lesions on the basis of commonly shared rules of evidence or inference. X number of Y cells per high power field are assigned different levels of significance in terms of indicating what constitutes cervical cancer. Such diagnostic assessments are commonly alleged to be unbiased and impersonal.

On this non-contextual view, it is as if various political, economic, social, and psychological differences among clinicians do not matter. Instead, pathologists and clinicians become placeholders for anonymous reasoners. The paradigm is that of a grand syllogism reasoned through the clinico-scientific community, viewed as the epistemic equivalent of the impartial, rational, fully informed observer. In terms of such a view, and given faith in the knowability of reality, a final and "true" diagnosis is sought, which should in principle be available to answer the question of why a certain clinical problem such as cervical cancer presents in a particular patient. Clinical diagnosis reflects, then, clinicians's epistemic interests [10] in "knowing truly" the character of clinical problems from the standpoint of an ideal observer so that this knowledge may be shared with others.

One might argue, however, that clinical diagnosis is more complex. The rules for acquiring evidence and drawing conclusions in the game of knowledge in general, and in medicine in particular ([6],[15],[4]), change over time. The ways in which we view clinical problems such as cancer depend on advancements in knowledge, in changing theoretical frameworks. One must, as a consequence, identify a diagnosis with a particular community, its rules for selecting evidence relevant to a diagnosis, and its rules for reasoning on the basis of such evidence. That is, clinical diagnosis becomes identified with particular clinical communities, understood as a group of individual clinicians who at a particular point in history share common rules of evidence and inference.

The ways in which we diagnose clinical problems such as cervical cancer depend, then, on a wide-range of value considerations. First, diagnosis depends on developments in the ways in which we *know* clinical problems. We know them through evolving clinico-scientific theories that are founded on standards regarding statistical relevance, coherence, simplicity, and elegance. Such standards appeal in part to a view regarding the nature of science. Such standards signify that a particular case is "worthy" of further attention and study. In this way, clinical diagnosis is value-ladened in a non-moral sense.

In addition, diagnosis depends on developments in the ways in which we manipulate and control clinical problems [14]. Both non-moral and moral values are found here. On the one hand, there is the concern to maximize benefits and minimize harms in clinical interaction. This is a non-moral consideration. Since diagnosis takes place over time, the logic of diagnosis must incorporate different transaction costs. There is a transaction cost when one moves from an initial presentation of the problem to one among a number of diagnoses at a later time. Transaction costs include a wide range of considerations such as morbidity costs, mortality costs, financial costs, and opportunity costs to the patient. (Costs presume, of course, a notion of benefit and an instrumental calculus.) Moreover, the order in which one acts to establish a diagnosis involves various costs to the patient. It is important to make a diagnosis fast when (1) a disease entails an increase in the probability of pain or death, and (2) there is an available treatment. In determining opportunity costs, then, one has to weigh the costs of the disease and the capacity to manipulate and control. As a result, there are major instrumental issues involved in deciding on any particular decision tree of flow pattern in making a diagnosis.

These considerations underlie a major debate in the clinical community regarding the ways in which clinicians classify carcinoma of the cervix. Figure 1 illustrates the relation between the traditional classification of abnormal Pap smears introduced in the 1940s and a new system increasingly adopted by medical laboratories in the 1980s. The system incorporates the cervical intraepithelial neoplasm (CIN) terminology to describe various degrees of preneoplastic lesion of the cervical and vaginal epithelium. The CIN system uses three grades of categories, CIN I, II, and III, which correspond to the descriptive terms "mild dysplasia", "moderate dysplasia", and "severe dysplasia-carcinoma *in situ*, respectively.

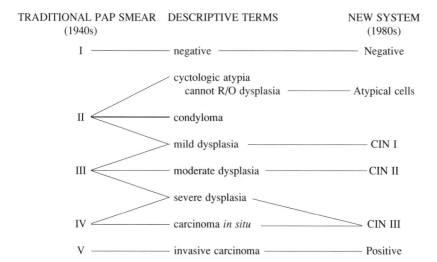

Fig. 1.   Classifying the Papanicolaou Smear. Under the new system, prescursors to squamous cell carcinoma are understood as part of a single disease continuum, what Richart (1980) calls cervical intraepithelial neoplasia (CIN).

The rationale for combining severe dysplasia and carcinoma *in situ* into one category is that (1) severe dusplasia and carcinoma *in situ* cannot be consistently differentiated morphologically, (2) there is no known difference in biological behavior between the two conditions, and (3) women with mild and severe dysplasia tended to be under-treated, and those with small carcinoma *in situ* tended to be over-treated.

As the third reason indicated, clinicians are concerned to minimize transactions (e.g., financial) and opportunity (e.g., morbidity and mortality) costs to the patient and related parties. If one adopts standards for treatment that are too lax, one may unduly increase the financial, social, and personal costs of care of individuals as well as society at large. However, if one sets the standards too strictly, one will pay the costs in loss of lives. As a consequence, one must decide a prudent balancing of the transaction and opportunity costs relating to over-and under-treatment. Such assessments require a prior determination of the comparative significance of the various possible benefits and harms involved in the various choices. Instrumental considerations are central to the ways in which clinical problems are understood and undertaken.

In addition to non-moral values, moral values influence clinical diagnosis. Ethical issues are involved for patients and physicians when any particular set of transaction costs and any particular pattern of transaction or intervention are chosen in making a diagnosis. How does one determine whether a medical intervention will do more good than harm, especially since patients may not endorse the same hierarchy of costs and benefits as do physicians? The acquisition of knowledge involves not only financial costs (e.g. how often is it worth doing a Pap Smear? [3], [8]), but exposures of patients to risks of morbidity and death.

In addition, diagnoses are rarely certain or even close to certain, so that in choosing among different diagnoses, one runs the risk of exposing patient to over-treatment and under-treatment. In order to manage these risks, special clinical classifications are created, such as the one by Richart that stages cancers in order better to coordinate treatment in terms of available information. The goal in such undertakings is to array actions with likely states of affairs.

Still the question remains: How does one determine how to set what weights to what values for which outcomes? At this point, one cannot avoid the moral or ethical issues at stake in clinical diagnosis. Ethical or moral issues involved in diagnosis provide the basis for blaming and praising in the moral community. A commitment to maximizing benefits over costs involves a commitment to beneficence, to maximizing benefits over burdens. The key here is commitment: once one commits to a goal, one engages in moral action. One is committing oneself to another in a particular way. In this case, a physician is committing his- or her-self to achieving the utilitarian goal of maximizing benefits and minimizing harms for a particular patient.

A commitment to a patient, such as "I will do my best for you.", is a commitment to respect persons, or to autonomy. Agreements or contracts are set up with patients to "do the very best" in terms of minimizing harm, discomfort, financial burdens, etc. Agreement provides the foundation for the moral action. Mutual agreement is the procedural expression of respect for person in the moral community. Through respect, one shows another that the self and the other are worthy of choosing, are worthy of being treated as responsible, free agents.

Ethical issues are involved not only for patients and clinicians, but for health care institutions, health care systems, and of course the public when clinicians choose a particular pattern of interventions and steps in making a diagnosis. Consider, for example, when a laboratory within a particular institution adopts one clinical classification for diagnosis as opposed to another. The decision will in part turn on what is taken to be the standard of clinical practice, on the institution's commitment to itself (e.g., in containing costs), and to its patients. Moral values turn in treating patients as persons, as individuals who own themselves (unless they have conveyed otherwise or lost authority to others).

CONCLUSION

The enterprise of diagnosis presumes a rich language for the description and explanation of clinical problems. Indeed, this richness has often caused confusion. In diagnosis, clinicians have hoped to discover without appeal to social values, those concerns that are properly medical and therefore properly placed within the language of illness and disease. This goal is, however, impossible to achieve. The domain of clinical diagnosis has historically been and remains a collage of various concerns to distinguish and resolve problems that are the result of physiological and psychological processes judged by individuals to be problematic. In this way, clinical diagnosis is contextual, embracing a wide range of concerns, interests, and goals of persons.

University of Colorado
Colorado Springs, Colorado

NOTE

[1] The Pap test or Pap Smear is a basic part of the standard gynecological examination for women. It is named after Dr. George Nicholas Papanicolaou (1883–1962), a Greek-born American physician who developed the test in the 1940s. The Pap smear is a means of detecting early cervical cancer. It is painless and without risk and is largely responsible for reducing deaths from cervical cancer by at least 70% in the last 40 years [1]. Recently, however, the test has come under criticism. Inaccurate cell retrieval and evaluation of the test report are two reasons given for the test failure. Richart, among others, goes further and criticizes the very way in which the classification scheme is fashioned [12].

REFERENCES

1. American Joint Committee for Cancer Stating and End Results Reporting (AJCCS): 1978, *Manual for Staging of Cancer*, AJCCS, Chicago.
2. Benson, Ralph C., ed.: 1984, *Current Obstetrics and Gynecologic Diagnosis and Treatment*, Lange Medical Publications, California.
3. Canadian Task Force: 1976, 'Cervical Cancer Screening Program', *Journal of the Canadian Medical Association* **114**, 1003–1033.
4. Cassileth, B.R.: 1983, 'The Evolution of Oncology', *Perspectives in Biology and Medicine* **26**, 362–374.
5. Engelhardt, H.T., Jr.: 1986, *Foundations of Bioethics*, Oxford University Press, New York.
6. Fleck, L.: 1935, *Entstehung und Entwicklung einer wissenchaftlichen Tatsache*, Benno Schwabe, Basel. English version, T.J. Trenn and R.K. Merton (eds.), F. Bradley and T.J. Trenn (trans.): 1979, *Genesis and Development of a Scientific Fact*, University of Chicago Press, Chicago.
7. Gross, R.: 1992, 'Intuition and Technology as the Bases of Medical Decision-Making', in C. Delkeskamp-Hayes and M.A.G. Cutter (eds.), *Science, Technology, and the Art of Medicine: European-American Dialogues*, Kluwer Academic Publishers, Dordrecht, Holland.
8. International Medical News Service: 1982, 'Less Frequent Paps Called Sheer Idiocy', *Ob. Gyn. News* (Dec 1–14), **4**.
9. Marx, J.L.: 1979, 'The Annual Pap Smear: An Idea Whose Time Has Gone?', *Science* **205**, 177–178.
10. McMullin, E.: 1986, 'Scientific Controversy and Its Termination', in H.T. Engelhardt, Jr. and A.L Caplan (eds.), *Scientific Controversies*, Cambridge University Press, Cambridge, pp. 49–91.
11. Portlock, C.S. and Goffinet, D.R.: 1980, *Manual of Clinical Problems in Oncology*, Little, Brown, and Company, Boston.
12. Richart, R.: 1980, 'The Patient With An Abnormal Pap Smear – Screening Techniques and Management', *New England Journal of Medicine* (February 7), 332– 333.
13. Schaffner, K.: 1992 'Problems in Computer Diagnosis' in this volume, pp. 197–241.
14. Weinstein, M.S.: 1983, 'Cost-Effective Priorities for Cancer Prevention', *Science* **221**, 17–23.
15. Wulff, H.T.: 1981, *Rational Diagnosis and Treatment*, 2nd ed., Blackwell, Oxford.

SECTION IV

COMPUTER AUGMENTED DIAGNOSIS

EDMUND D. PELLEGRINO

VALUE DESIDERATA IN THE LOGICAL
STRUCTURING OF COMPUTER DIAGNOSIS

I. INTRODUCTION: THE DECISION MAKIND TRIAD –
PHYSICIAN COMPUTER PATIENT

For centuries the healer in almost every culture was a mediator between
the sick person and the mysterious sources of healing. He was, so to
speak, between man and God – the friendly advocate, in touch, on the
one hand, with the pained body and spirit, and, on the other, with the
power from which healing could spring.

When medicine became scientific, the doctor was transformed into a
mediator with the quasi-magical powers of science. But as his science
grew, and took the form of technology, instruments interposed
themselves and began to distance the doctor and his patient. The doctor's
rounds in today's intensive care unit are at the foot of the bed – checking
monitors, reading reports from the distant laboratory or imaging
apparatus. The head of the bed where the person of the patient rests is
only too rarely visited. Dials are more often the recipient of the
physician's touch than the patient's body.

Still, the instruments and monitors seemed ostensibly to remain under
the physician's power because only he possessed the ultimate power to
diagnose, to choose treatments and to administer therapeutic procedures.
With the advent of the computer in diagnosis, in the selection of therapy,
and the evaluation of risks and benefits, the doctor's most sacred precinct
of power has been breached. He is pushed further from the patient, and
finds himself more between God and the machine, rather than between
God and the patient.

The physician-patient relationship is fast being converted from a
diadic to a triadic relationship in which the computer promises to be an
active, efficient, helpful, and not so silent partner. To what extent will the
physician be replaced by his mechanical partner? What realm of clinical
decision-making remains uniquely the physician's? How "objective" are
computer decisions?

These questions cannot be avoided by romantic rejection of the
machine, nor by its over-zealous adulation. There are plenty of

173

J. L. Peset and D. Gracia (eds.), The Ethics of Diagnosis, 173–195.
© 1992 by Kluwer Academic Publishers. Printed in the Netherlands.

protagonists for both extremes. But the ends of medicine and the patient's good can only be served by the most careful differentiation between what the computer does best and what only the physician can do. In his/her dialogue with both the patient and the computer the physician remains the value assesor – a role that cannot yet and perhaps never can be relinquished.

In this essay I wish to examine the essential features of the process of clinical judgment – selecting the right and good healing action for a particular person and, in the light of that end, to evaluate how physician and computer can best complement each other. Computer and physician both have roles. As with most other technology aimed at human service, we may not need *more* people, but we will need *more of* people.

Significant progress is being made both in modelling and simulating clinical diagnosis ([18],[9],[5],[14]). With the improved memory space and miniaturization of computers, and with better conceptualization of the logical structures of diagnosis, many of the current limitations of computers seem surmountable [17]. Clinicians will then comprehend better how they can interact with machines to make diagnosis safer, faster, more accurate, and more efficient. The advantages to patients of the proper use of computers will make their use morally mandatory.

Computers now monitor a wide range of physiological functions; they can control drug and fluid administration; implanted micro-computers can regulate cardiac rhythmns, release medication or hormones at selected sites, and assist in improving the hearing, sight, or muscular function of disabled persons. In addition to clinician-computer linkages, direct patient-computer interactions are also becoming a reality.

We can further envision a series of linkage that would progressively interpose machines between the physician and the patient. For example, a diagnostic-therapeutic system could be designed that starts with the patient giving his history directly to the computer. The computer could then choose which of a series of diagostic programs to enter, then by interaction with non-invasive imaging techniques like NMR, PET, ultrasound, and many new types of non-invasive diagnostic technology, the data of physical findings and history would be linked directly to computerized diagnostic and therapeutic algorithms. This whole series of patient-computer-computer linkages could end up with the diagnosis, the instruction to the patient on what to do next, whom to consult, what drug or diet to use, what side effects to watch for, when to check back, etc. The "electronic house call", with or without the clinicians's participation, and even robotic anesthesia and surgery of certain types controlled by computer, are not beyond imagination.

The logical limitations as well as the potentialities of such a scenario are many and are beginning to receive serious attention [19]. We need only mention the fundamental limitations of artificial intelligence systems – like the incompleteness of their knowledge base, the difficulties of expressing clinical knowledge in formalized language, of capturing the subtleties of the clinician's knowledge in rules, or his capacity to skirt rules when needed, and to encode new information and modify explanations in terms of the problem he faces, the inadequacy of computers in representing data dealing with the higher levels of biological organization and many others ([4], [27]). Clearly, the computer cannot replicate every subtlety of human behavior or the full context of clinical decision-making – nor is it likely to in the foreseeable future.

Beyond these limitations lies a more complex problem of a less technical nature. How do we decide whether to follow the recommendation, diagnosis, or descriptions of the computer? [36] The logic of a clinical diagnosis might be impeccable with, or without a computer, but that fact does not assure that a good decision will be made or a good action taken for a particular patient ([26], [27]). Despite these limitations, we can assume that the computer will be a partner in the physician-patient relationship aiding and even replacing the physician partly or completely.

The central question, then, in addition to the reliability of the internal logical processes of computer systems, is whether the results they produce in a particular case should be used and how. We deal here with questions of prudence, practical judgment, value and morals, with the covenant between physician and patient and with matters not transcribable into rule and language or susceptible to mathematical expression. These are qualitative judgments. All attempts to quantify such judgments suffer from a certain internal logical inconsistency.

It is precisely this aspect of the physician-patient relationship that ultimately determines where the physician fits in, where he can be replaced, and where he cannot. It is clear that the physician's role will change drastically in the coming era of man-machine relationships. These changes will become clearer if we examine the nature, and end of clinical judgment, and how value desiderata enter into the decisions to follow one course of action rather than another.

II. THE NATURE AND END OF CLINICAL JUDGMENT

Clinical diagnosis is one step in the larger context of clinical judgment. Clinical decision-making necessitates answers to four questions: What is wrong with *this* patient (*diagnosis*)? What can be expected to occur with this illness or combination of illnesses (*prognosis*)? What *can* be done to alter this projected course (*therapeutics*), and of the things that *can* be done, what things *ought* to be done (*clinical decision*)? [25].

When a patient consults a physician, the patient expects to be healed, to be moved from a state of illness to a state of health. All the interventions of medicine aim to effect this transition in states in a way which best serves the interests of a particular patient. The promise inhering in the "profession" of medicine is to take that action which is good for *this* patient. A good clinical decision must be scientifically correct, but it must also be "good", i.e., in the best interests of *this* patient. It must be in keeping with the patient's values, his vulnerable state as a sick person and his dignity as a being capable or reasoned choice.

The first step toward a right and good healing action for a particular patient is to place the patient as accurately as possible in a diagnostic category. Each category is defined by some finite constellation of symptoms, signs, and tests (SST) that carry with them a certain probabilistic relationship with the specific symptoms exhibited by this patient. The prognosis is likewise a probabilistic statement about the future course of this patient – the natural history of his illness deduced from the course of a universe of patients exhibiting as close a set of signs or symptoms as *this* patient. The diagnosis and prognosis, therefore, define the present state from which the patient must be moved if he is to be helped and the future course that must be altered. The patient hopes for cure, restoration to his original or a better state, and, if this is not possible, to be helped to cope with pain, disability, or dying if there is no treatment and the disease is terminal.

In the next step, the physician matches the most probable diagnostic category or categories in which his patient fits with the treatment interventions – drugs, operation, diet, environmental manipulation or talk, known to modify the natural history (expected future) in specific ways. A second probabilistic statement is then made for the way the natural history of each illness could be modified by each intervention for *this* patient in *this* particular stage of the disease, of *this* age, sex, occupation, genetic endowment, etc.

The third step is to select among the possible modes of management the one that is in the best interest, i.e., is most "worthwhile" for this patient. "Best interest" is a judgment encompassing multiple variables of a highly individual and personal nature – the probability of effectiveness of each treatment, the risk of side effects and toxicity, the degree, duration and kind of recovery possible and desired, the cost in dollars, time lost, disability and many other factors. Many of these factors can be programmed and computer assessed. But they must be given relative weights by the patient, who alone knows if what is proposed is worthwhile for him. Here the clinician confronts directly the patient's value system – that configuration of commitments and adaptations the patient has made up to this point in this life about what he deems "good", what he will pay, suffer, sacrifice, or die for. The patient's views of a good life and his utility assessment may differ sharply from those of the physician, of science, society, his family, or even of "reasonable men". Medical good is not the same as the patient's concept of good. Quality of life is not a medical determination. Only the patient can define what a satisfactory life consists in for him [27].

In the first two stages, scientific-actuarial reasoning operates more or less effectively and sophisticated computer systems can play a very significant role ([37], Fig. 1, p. 6). At every stage the good of the patient projects itself over each step from the beginning of data collection through the reasoning process, to the final selection of what *ought* to be done. Each step, therefore, is value-laden, and becomes progressively so as we approach the choice of the action that is "best" for this person.

It is precisely the sensitive appreciation of the simultaneous and continuous interplay of information and feedback between all the steps in clinical judgment that marks the best clinicians. Future sophisticated computer programs may encompass more of these simultaneous feedback loops. What they cannot reliably incorporate are the value desiderate that modulate the logic the physician and the patient use as they decide on an optimal course of action.

The clinician makes a diagnosis in order to treat – to cure, to care for, or to contain a disease in a human patient. The clinician's intrinsic purpose is a practical outcome, not scientific information about diseases. He needs to know what to expect of the patient's future course in order to know how to intervene and to modify that course in the patient's interests. For the clinician, in contrast to the biomedical scientist, diagnostic categories have an instrumental objective. The clinician needs a finite and manageable set of characteristics that will most accurately and efficiently locate his patient among several possible diagnostic, prognostic, and therapeutic categories. They must satisfy the criteria of utility more than of taxonomic completeness.

The clinician seeks, whenever possible, to approach diagnostic certitude. For this he needs pathognomonic observations like a biopsy, a bacteriological culture, a specific immunologic, enzymatic, or chemical determination. Often, when decision is most urgent these kinds of data are not yet known, or are not available for the disease in question. The clinician ends up more usually with several possible diagnostic categories, each supported by incomplete evidence of about equal weight. He is short of certitude even though the computer can assure him of a full display of the possibilities and the criteria required for each diagnostic category. He cannot make the final discrimination between disease patterns without some "weighting" of one piece of evidence over another.

Faced with these uncertainties, and the eventual need to taken an action, the clinician must invoke some principle by which to eliminate some pieces of evidence and include others. That principle is the good of the patient and it is expressed by an intuitive schema that enables the clinician to act prudently in the face of inadequate, uncertain, and often conflicting evidence. I have described some of these principles elsewhere as rules of clinical prudence [25]. These rules reflect certain values intrinsic to the clinician's function as a clinician. They are his personal choices of how best to fulfill his obligation always to act in the patient's best interests. They are based in his own interpretations of his clinical experiences, and are unquestionably influenced by his own "style" and values.

In whatever action he takes, the clinician seeks to maximize benefits and minimize dangers. He is, therefore, ultimately interested more in a *warrant for action* than in a satisfaction of all the criteria for a diagnostic category. The question the clinician asks himself is: What is the minimal, irreducible amount of information that makes a particular disorder likely enough to warrant treatment (action) based on the presumption of its presence? Rather than thinking in terms of matching the observations in his patient precisely with the full spectrum of SST characteristic or a disease pattern, the clinician is interested in that combination of characteristics that compel him to undertake treatment – to avoid the bad consequences of error, or inaction.

A patient with fever and a cloudy spinal fluid can with a high degree of certainty be diagnosed as having meningitis and that demands treatment. However, is the meningitis of bacterial, fungal, protozoal, viral, or mycobacterial origin? Each requires a different specific treatment. In the absence of identification of the organism, treatment must cover the most curable, and the most dangerous possibilities. Viral meningitides do not have specific treatments but the others do. If the evidence favors a viral meningitis but does not rule out the possibility of a bacterial meningitis, the patient should be treated as having bacterial meningitis. When there is any doubt the clinician must act in favor of the curable, serious, and treatable disease even if the weight of evidence is to the contrary. The "cost of error" is too large to be tolerated. Thus, the therapeutic trial becomes part of the decision schema.

The same would apply to pneumonias, poisonings, coma, chest pain, abdominal pain, headache, or in the psychopathological realm, to suicide risk, depression, homicidal tendencies. In each case, a computerized diagnostic system might list several plausible possibilities. The evidence might well point to one more strongly than the others. The experienced clinician however, will make his final choice among these possibilities on non-actuarial grounds by treating for the more dangerous, more treatable disease, leaving to time and the therapeutic test, the diagnosis of a non-treatable disease.

Acute abdominal pain illustrates this principle very well. A wide variety of abdominal lesions may be responsible – ruptured peptic ulcer, acute pancreatitis, small interstinal obstruction, appendicitis, rupture of abdominal aorta, mesenteric adenitis, conversion hysteria, strangulated internal hernia, twisted ovarian cyst, etc. Some of these entities demand immediate surgery; in some, surgery may be delayed, and in some it is positively contraindicated. Despite considerable improvement in diagnostic procedures, in many cases these possibilities are still sometimes difficult to distinguish pre-operatively. The surgeon must then look for the minimal signs that compel him to operate lest he miss a treatable and life-threatening lesion. He cannot wait for diagnostic certitude or closure; he may have to ignore ancillary data and computer produced assignments of probabilities and even the vaunted results of CAT scanning. Surgeons are excused a certain minimum of negative abdominal or MRI explorations, but *no cases* are excused in which life saving surgery is withheld.

Every branch of medicine embraces serious, yet curable, disorders.

Like the treatable acute abdominal emergency they must not be missed. The possibility of their presence dominates the clinician's choice of actions; they must be given weight out of proportion to their probability in a given case or population. Some curable lesions are relatively rare, like pheochromocytomas; some quite common, like parathyroid adenomas; some rare in certain locales and not others, like malaria, actinomycosis, or blastomucosis; still others like syphilis, gonorrhea or tuberculosis may crop up when we do not suspect them. Other diseases lurk where least expected, like tuberculosis, tetanus, botulism, diphtheria. They must be taken into account or the clinician will fail to treat a treatable disorder, even though its probability may be low. These diseases may be low on the statistical and stochastic scale.

Even when he is fully aware of the full set of criteria for a diagnosis, the clinician will often concentrate on data that will "rule-in" or "rule-out" the serious and treatable, or will given him warrant for action that will "cover" these possibilities even if the probabilities are strongly in favor of a less serious or a non-treatable disorder. Under these circumstances, he may resort to a therapeutic trial – "treating the treatable" – here a specific treatment is initiated and the diagnosis is made indirectly by observing the therapeutic effect.

Another way to assure that something important will not be missed is to maintained a high "index of suspicion". Here, instead of treating the probabilities, the clinician responds to the minimal criteria that suggest a serious diagnosis. Following through on these suspicions often uncovers a rare, esoteric disease, or a common disease manifested in an uncommon way. Most importantly, a high index of suspicion, i.e., clinical skepticism, will provide the necessary challenge without which the complacency of an established diagnosis will obscure a new or treatable disease in the same patient.

It is seductive to rest one's case in a dignosis that seems firmly established. Few things are more inhibiting to clinical skepticism than a diagnostic pigeon hole, particularly if it were fashioned by a prestigious colleague or institution. As a result the uncritical clinician can miss the primary diagnosis, the earliest signs of a new disorder, or the subtle changes in the progress of an old disease.

A large measure of skepticism is particularly pertinent to diagnoses that rest on X-ray, diagnostic image or laboratory data. Definitive as these can be, they may distract the clinician's attention from some other more important disease. Over-reliance on laboratory and X-ray procedures is the easy refuge, and the commonest intellectual error committed by today's physicians [13]. Though a particular test may suggest the high probability of a given diagnosis, the prudent clinician will not abandon the search for the treatable or the serious on that account alone.

As long ago as 1975 McNeil and her co-workers showed how the physician's concept of the patient's good determines the cut-off point between normality and abnormality in tests with a continuous scale of values [20]. The cut-off is paced at different points depending upon the cost of error in diagnosis. For a curable disease with high mortality, the physician seeks test of maximum sensitivity and tolerates lower specificity. For a disease with low mortality or no effective treatment, minimum sensitivity and maximum specificity are preferred. Similar considerations related to financial costs, risks, or discomfort of the procedure will further condition where cut-off-points are placed. When the costs of error are unknown, the points chosen will aim to minimize mistakes.

A clinician's principles of elimination reflect his personal clinical "style". By "style", I refer to the clinician's personal interpretation of sound clinical judgment, i.e., the summation of the logical and epistemological preferences that guide his decision-making. Style dictates the precise ways in which the normative aspects of diagnosis are modulated, or set aside, by individual clinicians.

For example, some clinicians's are compulsive diagnosticians and feel compelled to "prove" the diagnosis before acting, adhering strictly to the pre-established criteria [12]. They often add tests beyond the law of diminishing return. They cannot tolerate ambiguity or acting in the face of any uncertainty; they make a virtue of being "clinical scientists". Others may lean too far to the other side and rely on their intuitive grasp of a clinical situation remembering only their occasional triumphal guesses and never their more frequent dangerous conjectures. The obsessive clinician may over-value the computer in the interests of personal reassurance; the intuitive type may undervalue its enormous potentiality for completeness and orderly thinking.

Some seek diagnostic "elegance", analogous to the elegance of a physical theory. They look for the most economical synthesis of a clinical problem using just the right selection of observations and tests, those that are truly discriminating, none fewer and none more. The elegant diagnostician demands to know in advance how each test will advance the diagnosis or help in deciding on an action in the patient's behalf.

Therapeutic styles are just as variable. Some are therapeutic enthusiasts and believe in the pharmacological imperative. They see little harm in drug use and turn frequently to the prescription pad. Therapeutic enthusiasts rush to every new drug, and practice enthusiastic polypharmacy. The therapeutic trial becomes an agenda for a trial and error spree. Therapeutic enthusiasm aims to relieve every symptom; it fills the medicine cabinets of the world with its hit-and-miss nostrums.

Other clinicians are parsimonious therapists. They insist on demonstrable evidence of effectiveness for every agent. They usually disdain the use of multiple drug prescriptions, do not use placebos, resist chemical coping, and rely more heavily on patient education. They oppose the easy dictum that if a medication cannot do any harm, it "might do some good".

There are many variations on this theme of style – some clinicians are activists and have an incurable itch to intervene. They are the meddlers of medicine. Others prefer to permit a disorder to unfold with minimum interference, biding the proper time and choice of intervention. Some are cautious and fear harming the patient; others are willing to run risks. The range of clinical styles is as wide as the range of physicians' personalities.

Clearly each clinician brings an *a priori* set of values to the cost/benefit ratios at the branching points in a decision tree. Each clinician gives his own weight to the evidence and accepts a different warrant for taking an action. None of this should be taken as an argument against computerization. Computers can make the clinician's values and style more explicit, and their impact on his judgment more readily indentifiable. Those who design diagnostic systems have a "style" too, and this should not be forgotten or allowed to remain covert. They, too, must use principles of elimination and discrimination to simply decision-making. Without such "weighting", all data would be processed equally and the finer aspects of diagnosis – those most important to the patient – would be lost.

Every clinical decision should involve some participation by the patient and it must take into account his or her concept of what is worthwhile. Each patient exists in a particular frame of time, geography, personal history, and value preferences. They are part of his or her personal identity. They shape his or her part of his or her definitions of what is right, good and "valuable".

There are innumerable examples of marginal therapeutic endeavors that must be modulated by the patient's estimate of what is "worthwhile". Some patients will accept long, complicated, and risky treatment to extend their lives – others will not. Some medicalize every life event, implicitly trust their doctor's judgments, and submit to any procedure to attain "health". Others will want to restrict the use of medications and physicians, and will prefer prevention to cure; others will not want to be treated "at all costs" and will exercise a healthy skepticism about medicine and its practice. Some will want certain of their decisions to be handled by the physician and will not want to "participate" in clinical choices; others will insist on the highest possible degrees of autonomy.

Patient preferences must modify the conduct and conclusions of any diagnostic schema, computerized or not. Suffice it to say that every clinical decision involves an intersection of value systems – of the physician, patient, nurse, family, friends, colleagues, and society. Probabilistic logic provides the essential starting point for answering what is wrong, what a specific illness will do to the patient, and what treatment can do. But the decision about what *should* be done will evolve out of the accommodation each physician and patient makes with the value system of the other. Without such weightings, the finer adjustments of clinical judgment – those most crucial for the "right" and "good" choice for an individual patient – might easily be lost.

III. SOCIO-CULTURAL AND PSYCHOLOGICAL FACTORS

I have thus far emphasized the value desiderata that arise out of the nature and end of the process of clinical judgment itself. There are, needless to say, multiple sources prior to, and beyond the interstices of that process, from which clinicians and patients consciously or unconsciously, draw their value desiderata. Some are deeply rooted in historical, social, cultural, and ethnic contexts; others are structured into the way humans think, particularly under conditions of uncertainty. Only a few examples can be mentioned here.

Fleck in a seminal work has shown how concepts of disease and what constitutes a medical "fact" can change in different historical, social, and cultural contexts [11]. Engelhardt has underscored how concepts of good and evil determine what signs and symptoms we choose to include in a specific disease category [6]. He illustrates this graphically in his account of our changing ideas about masturbation as a disease [7]. Szasz has, as another example, criticized the criteria psychiatrists use to label the mentally ill [32]. Fabrega shows how the concepts of illness, therapeutics, and the sick role are conditioned by ethnic and cultural value systems [8].

Clearly, physicians and patients operate within multiple, intersecting value fields that form the framework for their judgments about what is good or worthwhile in a particular clinical decision. These fields powerfully modulate the logic of computerized diagnosis in subtle and, sometimes, covert ways.

Another important set of determinants of clinical thinking is linked to the nature of human thought itself. Tversky and Nisbet and their co-workers have examined how peoples' judgments of likelihood and probability depart from the laws of probability theory. Tversky shows how our preference for causal relationships and specific information leads us to ignore base-rate data. Those preferences lead also to what Tversky calls the "averaging error" – compounding causal data and averaging non-causal data – an error he found even in the evaluation of scientific data ([34], [35]). This habit violates the Bayesian rules for combining probabilistic data that operate independently of their possible causal interpretations.

Nisbet and Bogrida, in a series of experiments, attribute similar tendencies in estimating and weighing evidence to the relative vividness of specific instances over the abstractness of base-rate data [22]. In other experiments, Ceraso and Provitera show that errors in syllogistic reasoning are tied to the logical structure of the syllogism. Most were due to incomplete reasoning – interpreting the premises too narrowly or failing to examine the range of alternatives in a conclusion vigorously enough [3].

These habits of human thought operate in the design and interpretation of computerized diagnostic schemata as they do elsewhere. They are especially influential in the acceptance and use of the conclusions derived from the actuarial logic of computer diagnosis. It is important that they be identified and minimized, or at least accounted for in the evaluation of any proposed system of computer diagnosis.

Even more pertinent may be the logic of elimination by aspects that, as Tversky suggests, simplifies choices without reliance on estimates of relative weight or numerical computation [34]. Such a principle offers compelling reasons for clear-cut choices in any terrain in which the variables are many, complex, interacting, and conflicting. This is precisely the terrain of clinical judgments. The clinician's perception of the good of his individual patient and its projection over the whole process of decision-making does operate very much like a principle of elimination.

Tversky warns that elimination by aspects cannot be recommended generally because the choices eliminated may be better than those retained. Under specific conditions it can provide an approximation and a reasonable alternative to more complicated models. Covert principles of elimination, moreover, may influence computational models and dilute their supposed "objectivity".

But some such principle of elimination may be inescapable since there is so much uncertainty in assigning weight to the decision points in computational models. Their conclusion must be "weighted" before we can feel confident enough to put them into action on behalf of a given patient. However faithful a given diagnostic model may be to the logic of probability, the problem of the relevance of a statistical statement for an individual patient always remains. Algorithms, regression equations, the various forms of factor analysis, and decision trees improve our capacities to process information. But all the information they process is not of equal validity or utility for a good clinical decision. Rather, the ways in which we assign validity and utility to clinical information ultimately shapes the final choices of the therapeutic action that is taken.

The capacity to know when to modify, or ignore, the conclusions of a formal decision system separates the wise from the merely adequate clinician. We still lack reliable formal logical structures that can set forth the best set of alternatives for a given clinical situation and provide rules for choosing the best one among them. Not infrequently, precise scientific information (for example, the effectiveness or safety of a given medication or diagnostic procedure) necessary for an objective statement of probabilities is deficient as well. As Barnoon and Wolfe conclude, the feasibility and applicability of decision models to clinical decisions must overcome two major problems – one is the availability of reliable data in numerical form and the other the problem of reducing the dimensionality of a large set of attributes [1]. Yet the physician's focus on the optimal action to be taken in the face of multiple variable of uncertainty requires just this kind of information.

The wide range of the different kinds of knowledge that separate the personal physician epistemologically from the biomedical scientistis is very large [33]. Computational models cannot, as yet, encompass that full range that extends from the quantitative and observable to the qualitative, exhaustive, and personal. Blois' proposal of a vertical structure for medical rationing takes this into account though not completely (see conclusion). Principles of choice not encompassable in a computational model will therefore continue to operate both in the design and the interpretation of the conclusion of formal decision systems. The principle of "elimination by aspects" is as appropriate as any.

These modulations of logical formality by the habits of human thinkers are opeative in all decision-making. In clinical decisions they superimpose themselves like a grid over all the intricate steps in the process of clinical judgment. The key to the intrinsicality of clinical decisions is their *telos* – a technically correct and morally good healing decision for *this* patient, at *this* time, and in *this* clinical context. This telos acts in ways analogous to Tversky's principle of elimination.

IV. VALUE DESIDERATA IN SOME CURRENT MODELS OF CLINICAL DECISION

Most of the currently proposed models and algorithms and heuristic devices for making clinical decisions exhibit the value modulation and shaping I have been outlining. Schaffner has concisely analyzed the advantages and disadvantages of each type as logical structures [30]. Some indication of those points at which the value considerations I have outlined might shape different logical systems will further illustrate the major theme in this paper.

Systems using regression analysis attempt to describe the logic used in making predictions or decisions in statistical terms expressible in linear equations. This method, as Schaffner points out, is limited by the fact that the logic actually used is often curvilinear and configurational as well as linear. What is more, for precise application, linear analysis requires unidimensionality of data. One "dimension" of the data is the value question that cannot be expressed very well in numerical terms. Most crucial, perhaps, is the issue of data selection. When the diagnostician uses configurational logic, and when he decides what data to select, what variables to study, he will, consciously or not, resort to the values of "good medicine", "good" patient care, or what is "good" for this patient. These value considerations and the psychological factors that reflect the clinician's "style" can strongly influence the resultant regression equations.

Conditional probabilities have been central to current diagnostic models ever since the initial suggestion of the utility of Bayesian logic by Ledley and Lusted [16]. Schaffner empahsizes the dependence of Bayes' theorem on the independence of signs, symptoms, and tests associated with a disease. That this is not usually the case in actual diagnosis need not be overstressed here. It is enough to emphasize that the signs, symptoms, and tests the clinician chooses to enter into the Bayesian calculus are determined by the instrumental end they serve. Not only are the data not independent pathophysiologically but they are linked in a value context projected backward from the goal of doing what is good for *this* patient.

Bayesian systems work best when the data are limited yet still sufficient to generate reasonably valid conditional probabilities. In terms of this discussion, they would apply when the value considerations are small in number, clearly defined, and to some degree, independent of each other.

The diagnostic systems that most explicitly confront the problem of value modulations of logic are those accommodated by decision theory. These systems use logic trees with application of the Bayesian principles of conditional probabilities at each nodal point. At these points, quantitative cost/benefit assessment of outcomes of tests and treatments are introduced. The difficulties of Bayesian methodology *per se* are thus compounded by a number of other factors. Schaffner underscores the rigidity of these systems that makes modification and retracting of steps very difficult. Yet, the continual and simultaneous modifications between, and among, steps in the three stages of clinical judgment is an essential operation in the clinician's logical modus operandi.

Most important for the argument I am making in this paper is the assignment of utility coefficient cost/benefit (values expressed numerically) at the decisional branch points. This maneuver does confront value judgments directly and attempts to formalize them. But in assigning these value coefficients, the clinician's value preferences influence the outcome more directly than in any of the other proposed systems. This is so even in the more recent attempt to factor in the patients preferences. Moreover, the reasoning that results in an assignment of value may not be explicitly stated. As a result the uncritical clinician may grant the numerical expression of cost/benefit more objectivity than it can possess. These difficulties are readily apparent when models based on decision theory are applied even in such limited clinical problem as the work-up of hypertensive patients [23].

Several diagnostic systems have been devised to circumvent the difficulties inhering in strict Bayesian formality. They combine elements of categorical thinking with an approximation of Bayesian probabilistic logic. In general, they begin with a data base consisting of disease categories elaborated from disease manifestations. A particular patient's clinical manifestations are placed into several probable disease categories. These probabilities are condensed into those most probable, or plausible, by probabilistic rankings derived from the ability of each category to explain the patient systems. A variety of "weighting" maneuvers is then applied to reduce the possible diagnosis to one, or a few, most probable or plausible diagnoses.

Two systems have received considerable attention – MYCIN, which deals with the diagnosis of bacterial diseases, and CADUCEUS, which encompasses the majority of diseases encountered by the internist ([29], [31]). Both systems, despite their mathematical formalism, are susceptible to the intrusions of the clinician's value system as we have been outlining it in this paper.

The DIALOG system (now evolved as CADUCEUS) is an ambitious effort to simulate as closely as possible the actual reasoning modes of the clinician. The system starts with the development of an enormous data base constructed from most of the major diseases encountered in internal medicine. Disease categories are structured and substructured on the basis of common pathogenetic mechanisms or clinical manifestations. a complex diagnostic tree includes the hierarchical and associative relationships between manifestation and diseases. The data base is then further refined by a set of serial and feedback maneuvers too detailed to be recounted here. In the analysis of an actual clinical case, two phases are involved – the first is data entry that can further expand the data base if new manifestations are encountered, and an interrogative phase in which evidence for and against each evoked diagnosis in the diagnostic tree is weighted and subjected to a partitioning heuristic and then a variety of interrogative modes (RULE OUT, DISCRIMINATE, PURSUING). These operations are repeated until the evoked list is reduced to a conclusion that one or several diseases are present.

Several points in the logical maneuvers in development of the data base are susceptible to determination by *a priori* value assumptions of the kind we have been discussing.

One such point is the estimation of "evoking strength" on a scale of 0–5 from non-specificity to pathognomonicity of a test or manifestation. As indicated above, the cut-off point in any test is influenced by the instrumental end to be served by the test. A second point is in the TYPE classification, scaled 1–5, this time estimating cost and safety of the test in question. The third example is the IMPORT weighting, scaled 1–5 on the basis of how readily the manifestation can be ignored, presumably with cost and safety to the patient as determinants.

The MYCIN program works from a data base of some 300 rules that link specific observations with diagnoses [31]. The supporting evidence in different specific facts are combined by applying a "certainty factor" to each rule, i.e., a quantified estimate expressing the difference between measures of "belief" and disbelief". These maneuvers, as in the CADUCEUS systems, attempt to simulate the clinician's belief patterns. These certainty factors offer another example of points at which value considerations may strongly influence a formal logical system. "Belief" patterns are, at least in part, shaped by subjective attitudes about what is good medicine – for patients. Simply translating value-belief relationships into numerical terms does not eliminte the residuum of *a priori* values as influences in those estimates.

None of these considerations are to be construed as negating the very great value of efforts like MYCIN and CADUCEUS systems. They promise to enhance the clinician's capabilities in many ways if some of their complexities can be reduced to make them practicable for daily use. Indeed, by so explicitly simulating the clinician's reasoning modes, they help to localize the points at which logic and values intersect and to underscore the importance of an even better understanding of the value choices.

V. SOME IMPLICATIONS FOR MEDICINE

The availability of computer diagnosis and decision analysis need not be "dehumanizing" especially if patient preferences are included. Computers allow the clinician more time for the crucial third stage of clinical judgment – recommending and executing the right and good action for *this* patient. That end is best served if clinical diagnosis is as accurate, safe, efficient, and economical as possible. Everything that contributes to truly right and good clinical judgment makes it more beneficial for the patient. Projective schemata that infuse value considerations derived from the end of medicine into the whole of a formalized diagnostic system are essential to good clinical judgments. Computers by their special capacities to approximate human reasoning and to insure against its grosser errors are an invaluable asset. The physician has a moral obligation to use these methods whenever they are demonstrably helpful int he care of his patient. Such use is congruent with the traditional obligations of the physician to be both technically competent and personally compassionate, to make desisions, and to take actions that are technically and logically secure and, at the same time, morally good.

The physician has the further obligation to appreciate when he is replaceable and when means other than his own experience and intellect can better serve the interests of the patient. He need not worry about being replaced by the machine nor about its dehumanizing effects. Computerized diagnosis and decision analysis may or may not require fewer people. But like every new technological advance in clinical medicine, they demand more of people as people, more of their humanity, not less. Computer assisted decisions place the highest premium on those things only humans can do, i.e., empathizing, educating, caring for, making decisions with as well as for the patient.

VI. SOME IMPLICATIONS FOR THE FUTURE

(1) In the education of the physician and in his practice, less emphasis should be placed on storage of information in his/her brain and more on the optimal use of information sources, and how to evaluate the output of formalized diagnostic systems. A better knowledge of the logical and epistemic structure of computerized programs, decisions analysis artificial intelligence and the growing spectrum of information sciences is clearly indicated.

(2) The generalist can no longer be the versatile, all-purpose utility infielder expert at triage and little else. Rather, the generalist, must be adept in the specific intellectual functions needed to evaluate data outputs of all kinds, to select the appropriate program, in terms of patient good, and to decide when and how to enhance the patient's capacity for value choices. The generalist is simply irreplaceable in dealing with those aspects of care not reducible to rules and formalized language.

(3) Physicians will need a better education in the various forms of logic used in syllogistic, stochastic, and statistical reasoning. The limitations of those modes of logic as well as their advantages must be emphasized as well. In addition, the clinician now needs familiarity with the language, logic, and semantics of ethical analysis. Clinical decisions now explicitly demand a fusion of technical and moral endpoints.

(4) Research should, and must, continue in decision analysis, newer non-algorithmic heuristic modalities, and artificial intelligence. None of this need be dehumanizing unless clinicians allow it to be. This paper emphasizes the irreplaceability of the physician as a human being in every aspect of diagnosis and treatment.

(5) All of this enhances, rather than depreciates, the generalist. Indeed, the more specialized and narrow the diagnostic or therapeutic task, the more susceptible it is to formalized systems of analysis we have been describing. The more integration, weighing of values, and dealing with uncertainty are elements, the more is the generalist needed.

CONCLUSION

The machine and various types of mathematized diagnostic and therapeutic schemata are an integral part of the future of clinical practice. They can assist, but they can also complicate the management of individual cases. They are neither to be feared, nor worshipped. They are simply new instruments clinicians must add to their craft and link with their more traditional instruments for the betterment of patient care.

The healing relationship is no longer diadic. It has been polyhedral ever since physicians acknowledged seriously the participation of nurses, other health professionals, families, institutional representatives and society in what happens at the bedside. The new diagnostic systems, like the many persons now involved in decision-making must be integrated by someone who is the patient advocate and who has responsibility for the outcome. That person is still the physician. The challenge to education and practice is to relate all these elements harmoniously. To do so, physicians still must be guided by the most ancient of organizing, ordering, and justifying principles, the advancement of the patient's good.

Physicians can no longer be, as they were in ancient times, or even until the technological era, the sole mediators with the healing powers of God or the forces of nature. Physicians now must also be mediators with the machine, the mathematical schema, and/or the heuristic device, all of which accentuate rather than diminish the demands upon them as human beings.

In the last few years, the field of computer decision-making and particularly of decision analysis has progressed rapidly [24]. These developments support the line of argument pursued in this paper – namely that value desiderata enter at every crucial point and are difficult to quantify with any accuracy. This is especially true of the moral values which, I argue, must be factored into any decision that purports to aim at the patient's good [28] morally as well as technically.

Most notable of the recent advances is the acknowledgment of the importance of factoring the patient's utility assessments into decision analysis [15]. This is necessary for any morally valid decision process. The difficulty remains of quantifying "utility" which encompasses more than the algebraic sum of harms and goods, pleasures and displeasures. Utility, itself, is a value judgment, albeit one that represents the patient's rather than the physician's values.

Another recent development of interest is Feinstein's commendable effort to "harden up" to "soft" data, i.e., the clinical phenomena with which the clinic works daily [10]. His work focuses on providing a sounder theoretical basis for the construction and evaluation of clinical indices – a move which can be incorporated in computerized decision-making and decision-analysis schemata.

Note should also be taken of Blois' conceptualization of medical reasoning, description, and explanation as vertical processes. Blois proposes a hierarchical structure which extends from the molecular level of chemistry and physics, through cellular and organ levels of physiology, to the "patient as a whole" in clinical medicine. To this, I would add at the top an additional level of the moral and spiritual. Blois' delineation of multiple "levels" of complexity in medicine allows for different levels of decision analysis. Perhaps some day, long in the future, an attempt at an integration of all these levels into some grand decision-analysis tree may be possible. If we acknowledge the top of the hierarchical structure to be the moral and ethical, the importance of recognizing the value desiderate in clinical judgments, and the inherent difficulty difficulty of quantifying the qualitative will be further underscored.

Finally, mucn experience has been gained since the initial development of INTERNIST I [21]. Other schemata have been added like INTERNIST II, CADUCEUS, and specific programs for selected problems like chest pain, choices of antibiotics, etc. While precision, accuracy, and limitations of computer design are more clearly delineated, none of the reports to date alter the major line of argument I have followed in this paper for a better delineation of the value desiderata that enter all attempts to quantify and standardize clinical decisions, and the inherent difficulties to quantifying the qualitative.

Center for the Advanced Study of Ethics
and Center for Clinical Bioethics
Georgetown University
Washington, D.C. U.S.A

BIBLIOGRAPHY

1. Barnoon, S., and H. Wolfe: 1972, *Measuring the Effectiveness of Medical Diagnosis: An Operations Research Approach*, Charles C. Thomas, Springfield.
2. Blois, M.S.: 1983, 'Conceptual Issues in Computer-aided Diagnosis and the Hierarchical Nature of Medical Knowledge', *Journal of Medicine and Philosophy* **8** (1), 29–50.
3. Ceraso, J., and A. Provitera : 1971, 'Sources of Error in Syllogistic Reasoning,' Cognitive Psychology, 4, 400–410
4. Duda, R. O., and E.H. Shortlife 1983, 'Expert Systems Research', *Science* **220** (549), 261– 268.
5. Elstein, A.S.: 1976, 'Clinical Judgment: Phychiological Research and Medical Practice', *Science* **194**, 696–700.
6. Engelhardt, H.T., Jr.: 1976, 'Human Well-being and Medicine: Some Basic Value-Judgments in the Bio-medical Sciences', in Engelhardt, Jr. and Daniel Callahan (eds.), *Science, Ethics and Medicine*, H. Tristram Hastings-on-Hudson, N ew York, The Hastings, Center, pp. 120–139.
7. Engelhardt, H.T., Jr.: 1974, 'The Diseases of Masturbation Values and the Concept of Disease', *Bulletin of History and Medicine* **48**, 234–248.
8. Fabrega, H., Jr.: 1975, 'The Need for an Ethnomedical Science', in Charles Leslie (ed.) *Science Asian Medical Systems: A Comparative Study*, Berkeley, University of California Press.
9. Feinstein, A.: 1967, *Clinical Judgment*, Williams and Wilkins, Baltimore.
10. Feinstein, A.: 1987, Clinimetrics: New Challenges in Medical Measurement, Yale University Press, New Haven.
11. Fleck, L: 1979, *Genesis and Development of a Scientific Fact*, University of Chicago Press, Chicago.
12. Hardison, J.E.: 1979, 'Sounding Boards – to be Complete', *The New England Journal of Medicine* **300**(4), 193–194.
13. Harris, J.M., Jr.: 1981, 'The Hazards of Bedside Bayes', *Journal of the American Medical Association* **246**(22), 1602–2605.

14. Jacquez, J.A.: 1964, *The Diagnostic Process*, Ann Arbor, Michigan.
15. Kassirer, J.P.: 1983, 'Adding Insult to Inquiry: Usurping Patient's Prerogatives', *New England Journal of Medicine* **308**, 898–901.
16. Ledley, R.S. and L.B.: Lusted 1959, 'Reasoning Foundations of Medical Diagnosis', *Sciences* **130** (3366), 9–21.
17. Lewis, B.: 1979, 'Information: The Ultimate Frontier', *Science* **203**(12), 143–9.
18. Lusted, L.B.: 1968, *Introduction to Medical Decision Making*, Charles C. Thomas Publishers, Springfield, Ill.
19. McMullin, E.: 1983, 'Diagnosis by Computer', *Journal of Medicine and Philosophy* **8**(1), 5–27.
20. McNeil, B.J., *et al.*: 1975, 'Primer on Certain Elements of Medical Decision-making', *New England Journal of Medicine*, **293**(5), 211–215.
21. Miller, R.A., et al.: 1982, 'INTERNIST I, An Experimental Computer Based Diagnostic Consultant for General Internal Medicine', *New England Journal of Medicine* **307**, 468.
22. Nisbet, R.E., and E. Bordgida,: 1975, 'Attribution and the Psychology of Prediction', *Journal of Personality and Social Psychology* **32,** 932–943.
23. Pauker, S.G. et al. 1976, 'Towards the Simulatin of Clinical Cognition. Taking a Present Illness by Computer', *American Journal of Medicine* **60**, 981–996.
24. Pauker, S.F. and Kassiere, J.P.: 1987, 'Decision Analysis', *New England Journal of Medicine* **316**, 250 – 257.
25. Pellegrino, E.D.: 1979, 'Anatomy of Clinical Judgments: Some Notes on Right Reason and Right Action', H.T. Engelhardt, Jr., (eds.) S.F. Spicker and B. Towers *Clinical Judgment : A Critical Appraisal*, D. Reidel Publishing Co., Dordrecht, Holland, pp. 169–194.
26. Pellegrino, E.D.: 1979, 'Toward a Reconstruction of Medical Morality: The Primary of the Act of Profession and the Fact of Illness', *Journal of Medicine and Philosophy* **4**(1), 32–56.
27. Pellegrino, E. D.: 1983, 'Moral Choice, the Good of the Patient and the Patient's Good', in E.E Shelp (ed.), *Virtue and Medicine*, D. Reidel Publishing Co., Dordrecht, Holland, in press.
28. Pellegrino, E.D. and Thomasma, D.C.: 1987, *For the Patients Good: The Restoration of Beneficience in Health Care*, Oxford University Press, New York.
29. Pople, H.E., Jr., J.D. Myers, R.A. Miller,: 1975, 'Dialog: A Model of Diagnostic Logic for International Medicine', *in Advance Papers of the Fourth International Joint Conference on* Artificial Intelligence sponsored by the International National Joint Council on Artificial Intelligence, Tbilisi, Georgia, USSR, September 3–8, Vol. **2**, pp, 848–855.
30. Schaffner, K.F.: 1979, 'Problems in Computer Diagnosis', in J.L. Peset, D. Gracia, H.T. Engelhardt, Jr., and S.F Spicker, (eds.) *The Ethics of Diagnosis*, D. Reidel Publishing Co. Dordrecht, Holland, in press.
31. Shortliffe, E.H.: 1976, *Computer Based Medical Consultation*, Elseveir, New York.
32. Szasz, T.: 1961, *They Myth of Mental Illness*, Paul B. Hoeber, Inc, (Med. Div. if Harper Brothers), New York.
33. Toulmin, S.: 1976, 'On the Nature of the Physicians's Understanding', *Journal of Medicine and Philosophy* **1**(1), 32–50.

34. Tversky, A.: 1972, 'Elimination by Aspects: A Theory of Choice', *Psychological Review* **79**(4), 281–299.
35. Tversky, A. and D.: Kahneman D.: 1974, 'Judgment under uncertainty: Heuristics and Biases', *Science* **185**, 1124–1131.
36. Whitbeck, C. and R.: Brooks 1983, 'Criteria for Evaluating a Computer Aid to Clinical Reasoning', *Journal of Medicine and Philosphy* **8**(1), 51–65.
37. Wulff, H.: 1981, *Rational Diagnosis and Treatment,* Blackwell Scientific Publications, London.

KENNETH F. SCHAFFNER

PROBLEMS IN COMPUTER DIAGNOSIS**

I. INTRODUCTION

Computers have played experimental roles in clinical diagnosis for over twenty years. Interest in the possibilities of computer diagnosis was stimulated by the seminal article of Ledley and Lusted in 1959 [17] and research flourished in the 1960s. A number of conferences were held [13] [14] to facilitate interactions among investigators pursuing different approaches. In the 1970s, more sophisticated techniques of artificial intelligence (AI) were applied to computer diagnosis resulting in the development of a variety of programs possessing considerable conceptual depth. The enormous growth of medical knowledge continues to fuel interest in computer diagnosis as a means of obtaining a more manageable access to this knowledge.

In spite of the large number of projects and considerable funding support, computer diagnosis has yet to realize significant clinical acceptability. There are a number of reasons for this resistance, which stands in contrast to the use of computers in several other areas of health care delivery, including intensive care unit monitoring and adjustment of a number of therapeutic parameters in seriously ill patients (see, for example, Shortliffe ([32], pp. 3–12) and Groner et al. [9]). One class of problems for computer diagnosis is conceptual and it is this class on which I shall focus in this essay. I shall also mention, however, several social problems that will have to be resolved prior to acceptability of computer diagnosis in meaningful clinical environments.

Though what I characterize as conceptual problems can be roughly divided into two subclasses, which I shall term *logical problems* and *empirical problems*, these two subclasses are not fully independent and will be treated together in this paper. This interdependence arises since different logics of diagnosis may well require different types of empirical clinical information, and the availability or non-availability of clinical data can markedly influence the pursuit of different logics of diagnosis.

I shall not in this essay be discussing what the computer community terms "hardware" problems. In my view, the "hardware" or technological

197

J. L. Peset and D. Gracia (eds.), The Ethics of Diagnosis, 197–241.
© 1992 *by Kluwer Academic Publishers. Printed in the Netherlands.*

development of shared-time fast computers has progressed to the point where the major problems for computer diagnosis are not to be found here.

In the following section, I shall begin with a summary of both salient features as well as certain difficulties with two well-understood logical approaches to computer diagnosis, namely, Bayesian and branching logic programs. The following section discusses two elaborate but conceptually distinct diagnostic programs, the Stanford-based MYCIN program and the Pittsburgh-based INTERNIST/CADUCEUS program[1] and indicates what are some of their current problems. I shall conclude with a discussion of a number of desiderata for computer diagnosis programs and in that section touch briefly on some of the social issues regarding computer diagnosis.

II. BAYESIAN AND BRANCHING LOGIC DIAGNOSTIC PROGRAMS

Diagnosis of a patient's illness by a computer requires both clinical data concerning that patient as well as the use of a stored program. The program or "software" can be written in a variety of different computer languages such as MUMPS or LISP, but any program is itself the implementation of a more general logical approach that attempts to capture the essential cognitive elements of the diagnostic process. I am unaware of any in-principle reason why in some distant future a computer programmed with all known pathophysiological medical knowledge could not diagnose diseases in a way that might be quite different from the pattern of a contemporary expert clinician's reasoning. At present, however, such an approach is utopian and all of the general logical approaches to which I shall refer attempt to capture or model human clinical reasoning. This does not mean, however, that the models of diagnosis that are translated into computer language do not contain normative elements. In the several programs discussed in this article, we shall see the normative and the descriptive often intertwined.

A. Bayesian Approaches

My first class of logical approaches to modeling the clinical diagnostic process I characterize as Bayesian. There is an ambiguity in the use of this term: I shall reserve it for those reconstructions of diagnostic reasoning that utilize Bayes' Theorem to determine the posterior probability of a disease being present. Bayes' Theorem needs to be given certain information (to be discussed in more detail below) about prior probabilities of various diseases' occurrence and the probability, given the disease, of obtaining its signs, symptoms, and laboratory manifestations. In my next subsection, I shall discuss decision theoretic approaches that often enjoin the maximization of expected utility and that are associated with another sense of the Bayesian approach.

Bayes' Theorem is an optimal formula for revising our opinion in the light of prior opinions. It is usually construed as *normative* and psychologists have found that most individuals are more conservative than Bayes' Theorem licenses them to be.

Bayes' Theorem is a theorem that is relatively easy to prove using any standard axiomatization of the probability calculus. A considerable amount of research and debate in both statistics and in philosophy of science has gone into the problem of interpreting the notion of probability used in the Theorem (and in the probability calculus), and into analysis of the conditions under which the Theorem can be legitimately employed. The situation at the general level is still unsettled. In the area of our concern, namely, medical diagnosis, the early promise of using Bayes' Theorem to model clinical diagnosis, encouraged by Ledley and Lusted [17] and developed and tested via a number of Bayesian computer programs, has not been met. Bayes' Theorem still serves, however, as a basis for several ongoing research programs into diagnosis and also functions as an asymptotically limiting ideal in general. It thus deserves discussion.

Let us use the letter S to represent some diagnostically relevant evidence, say a symptom, a sign, or a laboratory value, (e.g., the white blood count). I will refer to S as a "manifestation". Then Bayes' Theorem can be stated in its simplest form as follows: the probability that a disease D is present given the evidence S, is equal to the product of the prior possibility of occurrence of that disease multiplied by probability of S given the disease, all divided by the probability of obtaining S in any circumstance, i.e., in the *general* population. This can be written as

$$P(D|S) = \frac{P(D) \cdot P(S|D)}{P(S)}$$

The Theorem is usually presented in a more general form. If there are n different diseases $D_1 \ldots D_n$ that can possibly cause S and these are mutually exclusive and exhaustive, we can represent the denominator $P(S)$ in a more explicit form as equal to;

$$P(D_1) \cdot P(S|D_1) + P(D_2) \cdot P(S|D_2) \cdot + \cdots + P(D_n) \cdot P(S|D_n)$$
$$= \sum_{i=1}^{n} P(D_i) \cdot P(S|D_i).$$

We can go even one step further and write a sequential form of Bayes' Theorem that allows one to continually update the probability of the disease given new information, say $S_1 \, S_2, \cdots S_k$, i.e., new signs, symptoms, and lab values. In this general sequential form, the Theorem is written as

$$P(D_i|S_1 \, \& \, S_2 \, \& \, \cdots \, \& \, S_k)$$
$$= \frac{P(D_i|S_1 \, \& \, S_2 \, \& \cdots \& \, S_{k-1}) \cdot P(S_k|D_i \, \& \, S_1 \, \& \, S_2 \, \& \cdots \& \, S_{k-1})}{\sum_{i=1}^{n} [P(D_i) \cdot P(S_k|D_i \, \& \, S_1 \, \& \, S_2 \, \& \, \cdots \, S_{k-1})]}$$

that tells us how the latest manifestation S_k affects the probability of Di assuming all earlier manifestations S_1 through S_{k-1}.

Bayes' Theorem in the form given thus far assumes that each evidential item or manifestation, say S_j and S_k are *conditionally independent*, i.e. that:

$$P(S_j|D_i) \cdot P(S_k|D_i) = P([S_j \cdot S_k]/D_i)$$

where $P([S_j \cdot S_k] D_i)$ would be a higher order probability (in this case a *second* order probability since the manifestations are taken pairwise). Note that what the independence requirement means is that given a disease, the signs and symptoms and laboratory values associated with the disease are independent. *This is a condition generally violated in medicine.* When the manifestations are conditionally dependent, additional information concerning the higher order probabilities is required that is often unavailable.

Some indication of the magnitude and seriousness of these problems can best be provided by outlining an example originally given by Szolovitz and Pauker [40] with some representative numbers. If we are considering a sequential series involving ten possible diseases and five diagnostic tests, say of the simply binary type which can yield a yes-or-no answer – and this is an unrealistically *simple* example, as we will see later – we need 63,300 conditional probabilities to cover all possible test histories. (The number of probabilities $= n \cdot \sum_{i=1}^{m} r^i \, [m!/ (m - i)!]$, for n diseases and i out of m possible tests, each with r possible results.) If, however, two diagnostic tests will yield the same results if we interchange their temporal order (and this is questionable in the case of some tests), we only need 2,420 conditional probabilities. If we now assume conditional *independence*, the number of conditional probabilities drops to $5 \cdot 10 \cdot 2 = 100$, which is just about manageable. However, as already mentioned the independence assumption is both implausible on pathophysiological grounds, and has been shown to be empirically false where it has been tested. (See, Jacquez and Norusis [15] and Szolovitz and Pauker ([40], p. 121).)

Furthermore, an additional assumption I have made, that the set of diseases can be partitioned into an exhaustive and mutually exclusive manner, is often violated. This can yield improperly normalized posterior probabilities in the case where exhaustiveness is violated. More serious is the exclusivity condition. Since few patients present with "classic" or completely typical forms of a disease and often have multiple diseases, probability methods require us to specify independent hypotheses *for every possible sub-type and combination of diseases*. This leads to a combinatorial problem of nightmarish proportions with an unmanageable demand for data, and also destroys the unifying view physicians take toward patients and their disease(s). (For further discussion of this, see Szolovitz and Pauker [40].)

This recitation of the problems with the Bayesian approach might lead us to ask why it is considered at all. First, in narrowly defined areas where the number of disease combinations is limited and where a good data-base is available to provide the conditional probabilities, the approach works well. (See Warner *et al.* [42]; and Jacquez and Norusis [15].) For example, Jacquez and Norusis [15] using an extensive amount of data provided by Pipberger and his colleagues drawn from the V.A. Cooperative study on Automatic Cardiovascular Processing [25] were able to obtain higher order conditional probabilities. They demonstrated, at least for this data, that higher order correlations were very significant in yielding correct diagnosis. (Interestingly, they were also able to obtain very suggestive results concerning the validity of the Bahadur approximation to provide complex conditional probabilities and also showed that clustering techniques work better than the (false) independence assumption or even actuarial methods on small data samples (n < 150).

B. Branching Logic

A number of distinguished clinicians, psychologists, and philosophers (Feinstein [3], [4], [5]; Kleinmuntz [16]; and Simon [37]) have argued that diagnostic reasoning is best understood as a branching or tree-like structure, in which each node represents a decision node or point of decision for the physician and which can have several paths leading from it to additional decision nodes. A simple example from Kleinmuntz [16] is provided as Figure 1 (p. 203) and a more complex example of a branching logic program is illustrated by the flow chart from Bleich's [1] acid-base and electrolyte disease program, given as Figure 2 (p. 204).

A branching structure can have "chance nodes" explicity added to it and then can be augmented by a decision theory such as a Bayesian analysis *with utilities*. At such chance nodes, then, the choice that was made would be, in the Bayesian decision-theoretic sense, dependent on the choice of the *maximum* expected utility, that is, the sum of the posterior probabilities multiplied by the utilities of each branch. Consideration of the "cost" and benefits of additional information could be entered. Such a Bayesian decision tree could also be (and has been) utilized for treatment choice as well.

Several investigators have examined such approaches. There are two particularly clear articles by Schwartz, Gorry, and their colleagues discussing diagnosis and treatment of severely hypertensive patients with possible functional renal artery stenosis ([30] [8]).

There are difficulties with this approach. First, the problems of obtaining the conditional probabilities afflict any Bayesian version of decision analyses. Secondly, the sequence of decision is extraordinarily rigid: once one has gone a way down the tree it is difficult to reverse the path and begin again. It is also difficult to modify and updata a diagnostic system with a tree structure. Interestingly, Schwartz, Gorry, and their colleagues have largely abandoned their decision-theoretic tree structure model in part for the reasons mentioned. Gorry [7] observed:

Not only did decision analysis fail to "fit" important aspects of complex clinical situations, but when the clinicians had good advice to offer concerning the real "trouble", the ability of the program to accept their advice was grossly limited by the necessary parametrization of their judgment... [G]generally we were unable to capture their approach to problems within our highly structured framework ([7], p. 25).

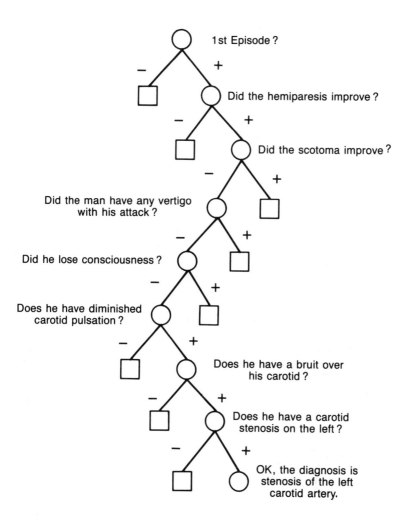

Fig. 1. A tree structure of a neurologist's diagnostic game in which the information given was: Sudden left central scotoma and right hemiparesis in a fifty-five year old. The process of diagnostic decision making may be visualized as a tree structure which represents the search strategy of a diagnostician. The circles represent nodes at which the physician may give a differential diagnosis, or he may elect to proceed by asking for more information. The plus or minus branches represent the presence or absence of the preceding symptom ([16], Fig. 1).

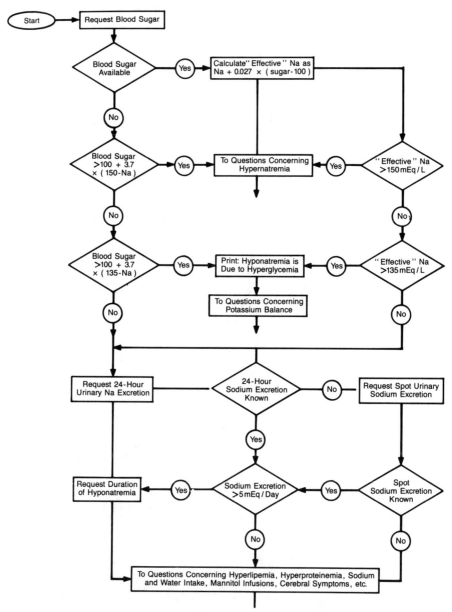

Fig. 2. Partial diagram of some of the pathways used to collect data concerning sodium balance. Equations for "Effective" Na and for blood sugar values that define significant changes in plasma osmolality are based upon a normal blood sugar concentration of 100 mg per cent, a molecular weight for glucose of 160 and an osmotic coefficient for sodium chloride of 1.8. Terminating arrows lead to pathways not shown in the figure; the arrow at the bottom leads to a series of questions dealing with edema, dehydration, congestive heart failure, Addison's disease, and renal disease, followed by questions concerning potassium balance ([1], p. 286).

III. COMPUTER DIAGNOSIS EMPLOYING
"SIMULATED" LOGICS

I believe that there are several lines of evidence indicating that what I term 'models of diagnosis employing a simulated logic' are likely to be the most promising approach. I think this will be the case both in the providing of a clearer understanding of clinical cognition as well as for physician users of computer consultations. A "simulated logic" is used here in the sense of an approximation to a probabilistic logic as represented in Bayes' Theorem. Each of the examples I will mention in this section utilizes a combination of categorial and probabilistic reasoning (following the terminology of Szolovitz and Pauker [40]). The probabilistic component of such a logic functions in a variety of ways from providing the degree of evidential support for a diagnosis in the MYCIN program, to generating the set of potentially explaining disease entities and then selecting a diagnosis in the INTERNIST-1 system. The probabilistic reasoning is involved in important comparative rankings of the ability of diseases to account for the patient's manifestations in PIP (the Present Illness Program of Pauker *et al.* [23]) and in INTERNIST-1, and is based on scoring numbers that are combined to produce totals using rules that mimic the intuitions of skilled clinicians. It is this mimicking aspect and the fact that no attempt is made, at this point in time, to force the reasoning into the Bayesian mode that suggests the term "simulated" logic.

Essentially identical remarks concerning "simulated logic" that appear in my article ([29], pp. 173–174) have been misunderstood by Whitbeck and Brooks [43]. Since the distinctions I have used are foundationally important, a few comments on their misinterpretation may be useful.

Whitbeck and Brooks ([43], p. 58) distinguish two senses of simulation, one that they attribute to Herbert Simon [36], and a "stronger sense" that they ascribe to me in my article [29]. They remark that I *contrast* programs employing simulated logic in my sense, essentially as given above, with programs that employ probabilistic reasoning. I did not and do not make such a distinction; rather, the contrast is between simulated programs and *Bayesian* programs – namely, those programs that conform to the strict canons of the probabilistic calculus and that employ Bayes' Theorem to modify belief in the light of new evidence. The distinction between Bayesian and probabilistic programs is important – the former are a restricted subset of the latter – but it appears to have been overlooked by Whitbeck and Brooks [43]. Whitbeck and Brooks also suggest that probabilistic systems "have an equal" claim to be called 'modeled' systems even though the authors of these systems are less likely to attend to existing approaches to clinical decision-making than are the designers of artificial intelligence systems" ([43], p. 58). I have no quarrel with this point since essentially the same view is proposed in the essay to which Whitbeck and Brooks refer, but it should be noted that the designers of the *Bayesian* systems have, on the basis of psychological literature, to construe their accounts as more normative than descriptive. Again, I believe that Whitbeck and Brooks' construal of my views is based on their failure to distinguish 'Bayesian' and 'probabilistic' notions.

In their argument against a stronger sense of 'simulated logic', Whitbeck and Brooks also, I think, overinterpret the remark they attribute to Shortliffe [32] namely, that some AI systems do "*not* reason as a human does". It may be helpful here to borrow on a distinction made in Hoffman [12] and cited in my article ([29], p. 164). One can conceive of a rough distinction between "paramorphic" models, which attempt to capture the same inputs and outputs as between humans and models, and "isomorphic" models , which attempt to represent portions of the *process* of reasoning interpolated between input and output. In MYCIN, Shortliffe and Buchanan developed the notion of certainty factors as a quantitative measure to represent the differences between experts' measures of belief and disbelief, about which I shall say more below. In their summary of their essay "A Model of Inexact Reasoning in Medicine," Shortliffe and Buchanan write:

This report proposes a quantification scheme which attempts to model the inexact reasoning processes of medical experts. The numerical conventions provide what is essentially an approximation to conditional probability, but offer advantages over Bayesian analyses when they are utilized in a rule-based computer diagnostic system. One such system, a clinical consultation program named MYCIN, is described in the context of the proposed model of inexact reasoning [33], p. 351).

On the basis of such remarks I continue to believe that AI programs are appropriately described as "simulated" in the isomorphic sense of mimicking but not exactly modeling human reasoning.

A. MYCIN

The MYCIN program's goal is to determine if a patient has an infection of the bacterial type, essentially bacteremias, and to recommend specific therapy. (Note that MYCIN is thus more than a diagnostic program.) The logic of the MYCIN program is complex. It employs about 500 rules of the "if-then", for example:

IF: (1) The stain of the organism is gram positive, *and*
    (2) The morphology of the organism is coccus, *and*
    (3) The growth conformation of the organism is chains

THEN: There is suggestive evidence (.7) that the identity of the organism is streptococcus.

These rules constitute MYCIN's knowledge base. The procedure is to use what is termed a "backward-chaining scheme" in reasoning to a diagnosis. In such a scheme, the data is entered or requested in the interactive mode by moving from the most specific and lowest points of a tree structure. This results in focussed questions to the user.

As alluded to above there is a complex methodology for combining evidential support provided by specific facts. Very briefly, there is a certainty factor (CF) associated with each rule that represents the degree of added belief attributed to the rules consequent given the truth of the antecedent. The certainty factor is the difference between two numbers: The measure of belief MB_H and the measure of disbelief MD_H, each of which are numbers between 0 and 1. There is a logic allegedly based on a fuzzy set strategy of passing belief values through AND/OR nodes to the consequent, which maximizes for "or" nodes and minimizes for "and" junctions.

Shortliffe and Buchanan [33] have provided a rationale for this non-standard logic that relies on some publications by Harré [11] and Törnebohm [41], and that has some similarities to Shafer's theory [31]. The logic, despite its extensive formal mathematical elaboration, has its source in an attempt to simulate the belief behaviors of clinicians, and is approximate.

Shortliffe [32] writes that

Diagnosis is not a deterministic process, and we believe it should be possible to develop quantification technique that approximates probability and Bayesian analysis, and that it is appropriate for use in those cases where formal analysis is difficult to achieve. The certainty factor model that we have introduced is such a scheme. It has been implemented as a central component of the MYCIN system ([32], p. 185).

The exact relationship of certainty factors to probability was unclear in its earlier stages, but additional work has introduced further clarification (see Pednault - I *et al.* [24] and Winston [44] pp. 191–199). One important feature of the MYCIN program is its explanation capacity. By examining the sequence of rules used, the program can answer questions concerning why a fact was used or how a fact was established. This strength, however, depends on the if-then rule format and is not easily obtained in other programs with a more flexible set of procedures.

The program is also easily modified and updated. If a physician objects to some reasoning, the program can lead him carefully through the decision process and can either result in the modification of a rule or the addition of a new rule.

The MYCIN program is quite successful in its domain of bacteremias and meningitis. However, this is a rather narrow area of focus and, as an example of computer diagnosis, it has the deficiency that the approximate diagnosis has to be made prior to computer consultation and the various culture tests ordered in order to answer the questions MYCIN asks. Questions may also be raised as to the general utility of the if-then or production rule format. As the number of infections MYCIN can diagnose increases, the antecedent component of the rules becomes more complex and unwieldy.

MYCIN has considerable value and has been shown to perform as well as or better than expert diagnosticians [45]. Clearly, when there is compelling evidence that bacteremias are often misdiagnosed and antibiotics grossly misprescribed [34], such a program has an important role to play. Whether it can be taken as a model for further computer diagnostic programs in view of the limitations of its production rule based format, however, is less clear. Interestingly, INTERNIST-1, the next example I shall discuss, began in its very early stages as a production rule program. This approach was abandoned, according to Myers (personal communication), as the rules began to "explode" [21] and it was subsequently based on a rather different logical format. It is to this that I now turn.

B. INTERNIST-1

The INTERNIST-CADUCEUS diagnostic system that utilizes AI techniques was developed by Jack Myers, M.D., a distinguished internist, and Harry Pople, a Ph.D. in computer science, beginning in 1974. Their work has gone through about four different phases of development. It is the second phase, in which the diagnostic program called INTERNIST-1 was developed, which is both the most well-known and the most empirically successful. In a very real sense, the INTERNIST-1 program is an AI simulation of Jack Myers's clinical reasoning processes and his own internal knowledge-base. (Myers is known to grimace mildly, however, when INTERNIST-1 is referred to as "Jack-in-the-box".) Randy Miller, M.D. soon joined the project, along with several programmers and (over the years) a number of medical students. The initial approach of Myers and Pople, which we might call phase 1, was toward extensive patho-physiological modeling and explanation. This turned out, relatively quickly, to be extremely costly and wasteful as a procedure, and in 1975 attention was redirected toward the type of program to be described in more detail below. It should be noted, however, that current research on the successor to INTERNIST-1, termed CADUCEUS, involves a return to an extent to a pathophysiological emphasis. I shall have more to say about this in a subsequent section.

The summary description of the INTERNIST-1 program I shall provide follows the article by Miller, Pople, and Myers [20]. Since I can be only brief in my description of the program, the reader is encouraged to consult their original article for details.

1. The INTERNIST-1 Program. An Overview of the INTERNIST-1 program should begin by mentioning its extensive knowledge-base. This comprises about 500 diseases known to internal medicine that are arranged in a disease hierarchy, i.e., from the general to the specific. The individual disease is the basic element in the knowledge-base. Each disease is characterized by a disease profile that is a list of manifesta-tions, i.e., signs, symptoms, and laboratory abnormalities, associated with

Table I. Interpretation of Evoking Strengths ([20], p. 469).

Evoking Strength	Interpretation
0	Nonspecific – manifestation occurs too commonly to be used to construct a differential diagnosis
1	Diagnosis is a rare or unusual cause of listed manifestation
2	Diagnosis causes a substantial minority of instances of listed manifestation
3	Diagnosis is the most common but not the overwhelming cause of listed manifestation
4	Diagnosis is the overwhelming cause of listed manifestation
5	Listed manifestation is pathognomonic for the diagnosis

that disease. Each manifestation in a disease profile has two clinical variables associated with it: an "evoking strength" and a "frequency". The
evoking strength is a rough measure on a scale of 0 to 5 of how strongly
does this manifestation suggest this disease as its cause. Table I above
contains an interpretation of these numbers.

The frequency, on the other hand, is a rough measure of how often
patients with the disease have that manifestation. This is measured on a 1
through 5 scale, an interpretation of which is given in Table II.

Table II. Interpretation of Frequency Values ([20], p. 470).

Frequency	Interpretation
1	Listed manifestation occurs rarely in the disease
2	Listed manifestation occurs in a substantial minority of cases of the disease
3	Listed manifestation occurs in roughly half the cases
4	Listed manifestation occurs in the substantial majority of cases
5	Listed manifestation occurs in essentially all cases – i.e., it is a prerequisite for the diagnosis

The crudeness of the scale is demanded because of the unreliability of
data in internal medicine. The medical literature is the source for frequency numbers, though again because of the quality of the data, a judgmental element figures in as well. In addition to these two numbers, each
manifestation is assigned an "import" on a scale of 1 though 5 as interpreted in Table III below. The import is a disease-independent measure
of the global importance of explaining that manifestation.

In addition to the disease profiles with their manifestations,
INTERNIST-1 also contains "links" between diseases, which are meant
to capture the degree to which one disease may cause or predispose to
another. There are also relations among manifestations as well, such as
"sex: female" is a precondition of "oligomenorrhea". Finally,
INTERNIST-1 contains a number of problem-solving algorithms that
operate on the individual patient data entered, using the information contained in the knowledge-base.

Table III. Interpretation of Import Values ([20], p. 470).

Import	Interpretation
1	Manifestation is usually unimportant, occurs commonly in normal persons, and is easily disregarded
2	Manifestation may be of importance, but can often be ignored; context is important
3	Manifestation is of medium importance, but may be an unreliable indicator of any specific disease
4	Manifestation is of high importance and can only rarely be disregarded as, for example, a false-positive result
5	Manifestation absolutely must be explained by one of the final diagnoses

As a specific case with the patient's manifestations is entered into the computer, the INTERNIST-1 program generates disease hypotheses that may account for that manifestation. This is a simple and direct triggering process that employs the lists contained in the knowledge-base linking manifestations to diseases. This set of disease hypotheses, which is usually large, is termed the master list. In addition, for each disease hypothesis, four associated lists are maintained that represent the match and lack of match between the specific patient under consideration and the disease profile. Each disease hypothesis on the master list is assigned a score on the basis of the match between the patient's set of manifestations and the knowledge-base disease profiles. Counting in favor of a specific disease hypothesis are the manifestations explained by that hypothesis. Credit is awarded based on the manifestations evoking strengths. Counting against a specific disease are (1) manifestations expected but found absent in the specific patient, which are debited in terms of the frequency values, and (2) manifestations not accounted for by the disease hypothesis, which are debited in accord with the import of that manifestation. In addition, a bonus is awarded to any disease that is related to a previously diagnosed disease via the links mentioned above. This bonus is equal to 20 times the frequency number associated with the disease in the diagnosed disease's profile.

A disease's total score is based on a non-linear weighting scheme that differs for scores based on evoking strengths, frequencies, and imports. Such non-linearity has often been found in descriptive studies of human judgment in a variety of disciplines, and thus should not be a surprising feature of INTERNIST-1. The weighting scheme assigns "points" in the following fashion:

(i) for evoking strengths: $0 = 1$, $1 = 4$, $2 = 10$, $3 = 20$, $4 = 40$, $5 = 80$

(ii) for frequencies: $1 = -1$, $2 = -4$, $3 = -7$, $4 = -15$, $5 = -30$

(iii) for imports: $1 = -2$, $2 = -6$, $3 = -10$, $4 = -10$, $5 = -40$,

in accord with the crediting/debiting procedure mentioned in the previous paragraph. These non-linear point assignments represent Myers's clinical judgment as honed by continuing experience with the INTERNIST-1 programs applied to patient cases.

All the disease hypotheses on the master list are scored in accordance with the procedure outlined in the two previous paragraphs. Then the topmost set of hypotheses above a threshold is processed further by a simple but powerful sorting heuristic that partitions this topmost set into natural competing sets of disease hypotheses. This allows INTERNIST-1, as Myers has noted, to compare "apples with apples and oranges with oranges". This sorting heuristic can be stated in various ways. One recent formulation is: "Two diseases are competitors if the items not explained by one disease are a subset of the items not explained by the other, otherwise they are alternatives (and may possibly coexist in the patient)" ([20], p. 471). This idea can be put another way be realizing that two different diseases, A and B, that meet this criterion will, if taken together, not explain any more of the manifestations than either one does taken alone. This sorting heuristic thus creates a "current problem area" consisting of the highest ranked disease hypothesis and its competitors. Miller, Pople, and Myers [20] refer to this procedure for defining a problem area as "ad hoc", and note that because of the procedure, INTERNIST-1's "differential diagnoses will not always resemble those constructed by clinicians" ([20], p. 471). This, it will turn out, is a critical difficulty for the program, and a point to which I shall return to below.

At this point INTERNIST-1 will either conclude with a diagnosis or will commence with one of its interactive searching modes. A diagnosis is concluded if the leading disease hypothesis is 90 points higher than its nearest competitor. This value was selected because it is just a bit more than the "absolute" weight assigned to a pathognomonic manifestation. It should be stressed here that this method of concluding a diagnosis is comparative with respect to its competitors as defined by the sorting heuristic, and is not based on any absolute probabilistic threshold, such as > 0.9. If a diagnosis is not concludable, INTERNIST-1 enters one of its searching modes.

In a searching mode, additional information is requested from the user. The specific mode that is entered, and thus the type of question posed, is determined by the number of competitors within 45 points of the leading diagnosis. If there is none, then INTERNIST-1 enters the "pursuing" mode, posing specific questions to the user that have a high evoking strength for the topmost hypothesis. If the answer is yes to these questions, the program will rapidly reach a concludion. If, however, there are a number of competing hypotheses running neck-in-neck, specifically 5 or more within 45 points of the leading diagnosis, the program enters its "ruleout" mode. Here questions that have high frequency values for the competitors are asked of the user, the rationale being that negative answers will result in the rapid elimination of some of the contending hypotheses. Finally, if 2 to 4 hypotheses cluster within 45 points of the leading diagnosis, the program enters its "discriminating" mode. Here questions are posed to the user that are likely to yield answers that would increase the separation in scores between or among the competing hypotheses.

An examination of patient cases in the literature (see, for example, Miller, Pople and Myers [20] or Schaffner [29]) will disclose that several questions are asked at a time, and that the program then recalculates its scores; it may well also repartition its diagnostic hypotheses. Questioning proceeds from the more easily obtained data to the more expensive and invasive information. Once a diagnosis is concluded, the manifestations that are accounted for by that diagnosis are removed from further consideration. This is an important point that should be noted; I will comment on it again in connection with some of its deleterious consequences. If there are manifestations that are not accounted for, the program recycles and by this process can diagnose multiple coexisting diseases.

There are two circumstances in which INTERNIST-1 will terminate without reaching a diagnosis. First, if all questions have been exhausted and a conclusion has not been reached, the program will "defer" and terminate with a differential diagnosis in which the competitors are displayed and ranked in descending order. Second, if all the unexplained manifestations have an import of 2 or less, the program stops.

INTERNIST-1 is a powerful program, surprisingly so in the light of its *prima facie* 'brute force' methods of analyzing clinical data. A formal evaluation of its diagnostic prowess by Miller, Pople, and Myers [20] indicates that its performance is "qualitatively similar to that of the hospital clinicians but inferior to that of the [expert] case discussants". The program does, however, possesses a number of problems on which work continues to be done. These problems are pointed out by Miller, Pople, and Myers [20] as well as by Pople ([27], [28]).

C. An Example from INTERNIST-1

It will be useful to review a fairly difficult clinicopathologic conference (CPC) case, this one from the May 18, 1978 issue of the *New England Journal of Medicine* (Appendix I). The case is reasonably representative in the sense that the performance of INTERNIST-1 on this case is roughly typical of its operation (Appendix II) and the parallelism between the clinicians' reasoning and INTERNIST-I's analysis is striking.

In the first abstraction of the data from the case, there was only enough data from the patient's first two admissions for INTERNIST-1 to begin in the *ruleout* mode, with Systemic Hodgkins disease as its top candidate in addition to listing four other diseases. Additional data available to clinicians (which was not initially entered due to an abstractor's oversight) led the program immediately into the *pursuing* mode, focussing on Hodgkins disease. The clinicians at Massachusetts General Hospital's conference did not diagnose the patient's illness (which was Hodgkins diseases) quite correctly and in the pathological discussion it was noted that "the hepatic infiltrates of Hodgkins disease may vary in cellular composition...and that the involvement may stimulate benign granulomatous disease" ([18], p. 1137). The pathologist also added that "the diagnostic Reed-Sternberg cells *in the bone marrow* [biopsy] provided the first clue that we were dealing with Hodgkins disease rather than a benign disseminated granulomatous disorder" ([18], pp. 1137–1138). (My emphasis.)

We ran several variants of this case on INTERNIST-1. The explicit *denial* of Reed-Sternberg cells in the biopsy would send INTERNIST-1 off the track and ultimately into a *deferring* mode in which it would only produce a differential diagnosis containing Hepatic Hodgkins and Hepatic Sarcoidosis in that order. (Sarcoidosis was one of the diseases entertained by the clinicians.) If, on the other hand, the *explicit denial* of the Reed-Sternberg cells were *reversed* to an affirmative, as was discovered in the bone-marrow biopsy, INTERNIST-1 concluded with

Hodgkins. Most interesting, if the explicit *denial* were withheld by entering a "not available" (N/A), INTERNIST-1 was able to conclude Hodgkins. Clearly the program is sensitive (and conservative), but it appears to model reasonably closely a clinician's reasoning process.

D. Critiques of INTERNIST-1; INTERNIST-2 as an 'Unstable Intermediate'

In spite of the striking performance of INTERNIST-1 noted in the previous sections, its developers have felt for some time now that fundamental further refinements were needed in the program. The paper by Miller, Pople and Myers [20] and other papers by Pople ([27], [28]) provide critiques of the program. One difficulty is pragmatic but none the less very important, namely, the inappropriate initial focus of INTERNIST-1 on complex problems. As Pople notes ([27], p. 148), this rarely leads to false conclusions, but the amount of time the user is required to spend at the terminal and the number of questions asked would lead to user nonacceptance. This seems to be a consequence of the scoring procedure that provides higher scores for unifying disease hypotheses even if two distinct hypotheses might both account for the data better and lead to a diagnostic conclusion faster. This intuition set the stage for a temporary successor system known as INTERNIST-2, on which I shall comment below.

Miller, Pople, and Myers [20] also note that INTERNIST-1 cannot reason causally (its link structure excepted) and accordingly cannot provide appropriate explanations. In addition, it does not handle the interdependency of manifestations well, which probably will require a more causally structured knowledge-base for a solution. The program is also deficient in that it has no anatomical knowledge and "cannot recognize sub-components of an illness such as specific organ system involvements or the degree of severity of pathologic processes" ([20], p. 474).

An additional criticism relates to the *serial processing of a problem area generated by the existing scoring algorithm and partitioning heuristic.* A careful examination of the italicized terms contained in the previous sentence will indicate how close to the heart of INTERNIST-1 this criticism lies. At least two attempts have been made to deal with this problem, which has several different facets. Pople [28] discusses a short-lived successor to INTERNIST-1 that was termed INTERNIST-2. This program, which can be termed phase 3 of the INTERNIST group's research program, was implememented and dealt with some complex cases rather well. Some specific problems developed in that implementation, however, which set the stage for the current and final fourth phase of development represented by CADUCEUS. A brief description of INTERNIST-2 will be useful to us since portions of its approach have been generalized in CADUCEUS.

INTERNIST-2's primary goal was to develop a methodology for concurrent problem formulation. To this end, Pople introduced several new concepts into the INTERNIST-1 program. First, a notion of a 'constrictor' was developed as a generalization of the idea of a pathognomonic manifestation. A constrictor was invoked as a tool for generating multiple hypotheses. Pople wrote:

> If the focus of attention is directed at higher levels of the disease hierarchy rather than at the terminal level nodes, quite specific associations between very commonplace manifestations and these higher level disease descriptors can often be established. For example... bloody sputum, while not pathognomonic with respect to any particular lung problem, provides ample justification for serious consideration of the lung area as a problem focus ([28], p. 84).

Second, Pople used this notion of a constrictor to develop a multi-problem generator involving a modified scoring algorithm. The technical details cannot be presented in this paper but can be found in Pople's forthcoming paper. Third, Pople introduced a procedure for handling a combination of two (or more) interdependent disease hypotheses as a unitary hypothesis. Terming a single disease solution to a patient's illness a simplex, Pople defined a synthesis operator that:

> maps partially expanded hypothesis states into new states wherein two or more simplex nodes have been combined into a single complex node. The basis for such combinations is the existence of causal, temporal, or other known patterns of association among members of these simplex problem sets ([28], p. 85).

The new strategy using these new notions of constrictor, multi-problem generator, and synthesis operator appeared quite promising and work on this approach to a successor to INTERNIST-1 proceeded for about another year after its initial implementation. Unfortunately, this task ultimately did not prove fruitful. In his commentary on INTERNIST-2, Suppe [38] cites personal communication with Pople indicating that the heuristics that are described in Pople's paper on INTERNIST-2 [28] were not sufficiently powerful, and that "in particular they "washed out" various intermediate links such as the notion of infection which in actual medical practice play an important role in the clinical diagnostic procedure". The attempt, nonetheless, was a significant one, and constitutes an important, if unstable, intermediate (to use the chemists' language) between INTERNIST-1 and the more elaborate successor known as CADUCEUS.

E. Parallel Processing Tangled Hierarchical Classification and CADUCEUS

INTERNIST-2 represented an attempt to develop a concurrent problem formulation approach to clinical diagnosis. Interestingly, various tries at implementing this notion of concurrency suggested to Pople that a fundamentally different conceptualization of problem-solving was likely at work here.

In his essay on CADUCEUS, Pople [27] begins by stressing that the task of AI in medicine may well have been misconceived by many of its practitioners. Using a term developed by Simon [35], Pople suggests that medical diagnosis is better conceptualized as an "ill-structured problem". Many AI programs seem to assume that diagnosis occurs within the context of a well-structured problem, i.e., *after* we have restricted our attention to a differential diagnosis. A, if not *the*, major difficulty for Pople, is to convert an ill-structured problem into a well-structured problem. Problem formulation, and in particular problem synthesis, is the key task. Pople argues that this process will require the most sophisticated representational and heuristic techniques available in the AI armamentarium. As Pople writes:

A particularly important aspect of the overall design is what in the terminology of artificial intelligence programming is referred to as a "control regime", or strategy by which the focus of attention of the problem solver is first directed at one alternative and then another, perhaps with resumption for consideration of those options initially rejected. In situations that admit of alternative conceptualizations, there must be some machinery to generate the space of feasible task formulations, and to manage the "shift of paradigm" that takes place when moving from one such conceptualization to another. Moreover, criteria must be developed by which to judge the various formulations of the task so as to guide the search process towards that task characterization which is actually correct. In addition, there must be some means of determining when this metagoal of the problem formulation process has been achieved ([27], p. 129).

I believe we can see the rationale for this new stress in CADUCEUS if we recall the problems with INTERNIST-1. (Remember that

INTERNIST-2 was developed to deal with the difficulty of an overly-focussed, serially-processed problem area. Recall the ad hoc construction of the differential diagnosis that was so central to INTERNIST-1's success as well as its difficulties.) Concurrent problem formation in INTERNIST-2 was promising, but attempts at implementing this notion ultimately led Pople to realize that a modification of the INTERNIST-1 knowledge-base would be required. Some brief comments on this reorganization and the need for it will be useful.

Pople remarks that though it was initially hoped that "the necessary constrictor patterns could be extracted from the original INTERNIST-1 knowledge structure using heuristic mapping rules" [27] this, unfortunately, did not turn out to be the case. Pople notes that there are two reasons for this:

use of a strict hierarchy in INTERNIST-1 for defining categories of disease which has served to make diffuse what would otherwise be strong constrictor relations, and elimination of pathophysiological detail which cannot be restored without a considerable investment of medical expertise. Both of these call for major revision to the structure of medical knowledge underlying the INTERNIST program ([27], p. 151).

Pople's approach is to develop a "synergistic blend" of a causal or patho-physiological model with a nosological structure. Pople proposes that a strict hierarchical classification is "too restrictive" and suggests that an alternative taxonomy using the notion of 'organ system involvement' may be more appropriate. This will be a more complex typology "in which any given disease can be classified in as many descriptive categories of the nosology as are appropriate". Pople adds that "this type of acyclic graph structure, which allows any given node to have an arbitrary number of parent nodes, is sometimes referred to as a 'tangled heirarchy'" ([27], p. 157).

Implementing this reclassification requires several new notions which can be mentioned only briefly. A more extensive causal net must, of course, be introduced tying diseases and manifestations together. The added detail, however, may obscure a less detailed but unifying common cause. This would be unproblematic if the entire net were simultaneously processable, but traversing and probing such a net in a variety of directions is time-consuming and costly. Accordingly, some new heuristic must be introduced that will facilitate searching such a network. Pople proposes a new type of link referred to as a "generalized link" or "planning link" that he believes will enable "formulation of parsimonious refined differential diagnoses – in many cases permitting a single category to be selected as the scope of a manifestation (in which case the relationship is that of constrictor)" ([27], p. 159).

Just as the constrictor concept is generalized in his article [27] to the more complex notion of a planning link, so Pople develops a generalization of the synthesis operator introduced in this discussion of INTERNIST-2. In CADUCEUS, some six different simple synthesis operators are proposed that can be utilized in combination to synthesize various subtasks. The combination of these various simple operators together with planning links and their subpath instantiations constitute a "generalized multistep unification operator". These operators permit the synergistic blending of causal net focussing with nosological focussing. Considerable detail in these operators and their application are to be found in Pople's article [27], to which the reader must be referred. It should be pointed out, however, that there is as yet in CADUCEUS no simple and straightforward algorithmic procedure for achieving a final synthesis. A number of hints exist in accord with what was described above. Further work within the classical AI paradigm of "state search space" is needed (see Nilsson [22], for examples). This work will need to be supplemented by a yet to be developed "high level control program", which would implement Simon's approach to solutions of ill-structured problems.

Pople summarizes his expanded and reorganized vision of medical diagnosis and the place of CADUCEUS within a reconceptualized task domain as follows:

This new knowledge representation [in CADUCEUS] provides multiple nosologic structures, by which disease entities may be classified in as many descriptive ways as appropriate. In addition, there is provision for a representation of detailed pathophysiology, by means of a causal graph having no restriction as to level of resolution. These basic structures are supplemented by generalized links – a subset of the transitive closure of the causal graph – which provides for as rapid convergence on tentative unifying hypotheses as in INTERNIST-1, while at the same time enabling access – via a sub-path initiation mechanism – to as much detail as is available in the underlying causal graph. This underlying result has been facilitated by means of a path unification algorithm used to combine elementary task definitions into unified complexes. As application of this synthesis operator cannot be considered irrevocable, it is necessary to envelop these heuristic maneuvers within a sophisticated control regime. Thus we have discovered within the task environment of medical diagnosis a core problem, the solution of which requires some of the most powerful methods available in the armamentarium of artificial intelligence ([27], pp. 183–184).

I believe that this rather lengthy account of the development of the INTERNIST/CADUCEUS research program indicates both how far investigators have proceeded in capturing the nature of clinical reasoning, and also the complexity and the enormity of the task still to be accomplished. The CADUCEUS program is still not operational, but even at this stage of its implementation it suggests through its own successes and difficulties, various further directions for development. Quite recently the original INTERNIST-1 program has undergone further augmentation and has been implemented in a form which can be purchased by practicing physicians and run on a desktop personal computer. This version is now known as the Quick Medical Reference or QMR, and is described in [21].

This brief discussion of INTERNIST-1, INTERNIST-2, and CADUCEUS suggests some of the problems that are associated with this diagnostic program. There are other features contained in an expert clinician's reasoning that are still missing from CADUCEUS. For example, utilities (or disutilities) are not built into the data base except via stages in information requested, which move from inexpensive and non-invasive data to more costly and more dangerous procedures. A clinician may well focus first on a potentially serious disease that must be ruled out even though unlikely. Whether a disease is treatable may also figure in the diagnostic process, and if further differentiation is not useful for treatment, diagnostic pursuit may be terminated. These considerations are part neither of the CADUCEUS program nor its ancestors. The program also has a tendency to continue working on left over or unexplained findings and requesting additional information. A clinician can often recognize this remainder as irrelevant or a consequence of an atypical form of diagnosed disease, but INTERNIST-1 often does not, though the stopping rules mentioned earlier may lead to termination. This is not a serious problem since the program is intended as a consultation aid to a clinician and not a decision-maker, but it does not illustrate the additional flexibility of the physician.

Because INTERNIST-1 utilizes a "comparative" weighing, this remainder, if not constrained by another competing disease, can occasionally generate a diagnostic conclusion which is inappropriate. This is in part a consequence of the currently incomplete data base and it will be interesting to see if these "bugs" disappear when the knowledge-base is completed.

Finally, none of the programs discussed have any sense of time – no capacity to appreciate a disease's development or the capacity to defer until further information is available at a later time. Successive admissions can be entered by citing some earlier information as history items, but this often does not capture the information one wants to encode. How serious this problem is, and the extent to which it can be outflanked by CADUCEUS, remain undecided.

IV. DESIDERATA FOR COMPUTER DIAGNOSIS

In this section, I am going to bring together portions of the earlier criticisms and attempt to sketch what appear to be reasonable desiderata for computer-based diagnostic programs. It must be stressed that these represent ideals that will only be met by extensive (and expensive) research by a variety of groups over a considerable amount of time. I am going to group the desiderata into "logical" (but including some empirical components) and "social", for reasons that I develop below.

A. Logical (and Some Empirical) Desiderata)

In order for a computer diagnostic system to operate accurately, it requires (1) *an accurate knowledge-base.* Such a knowledge-base is encoded in MYCIN's production rules and in the INTERNIST/CADUCEUS disease hierarchy. Occasionally, mistakes or additions are noted even in "completed" disease profiles, which leads to my second desideratum, (2) *an easily modifiable and updatable knowledge-base.* This desideratum was touched on in connection with problems afflicting branching logic diagnostic systems where such modifications are difficult to incorporate because of their effect on other parts of the system. I have also argued for (3) *an initially non-Bayesian* approach except in certain well-defined sub-areas where the extensive and complex probablistic data is available, and for (4) *a simulated or modeling approach* to diagnosis. I have also suggested that none of the programs of which I am aware are complete, and it may also be that there are covert generalizations concerning clinical reasoning that have yet to be captured even by the most sophisticated programs. Accordingly, I would add to the desiderata the need for (5) *a sensitivity and openness to alternative logical structuring of a program,* such as was discussed above in connection with the transition from INTERNIST-1 to its successors, with the realization that such shifts are likely to be difficult and expensive. Simulated programs also currently require (6) *further normative logical development* perhaps under asymptotic Bayesian constrainsts, but perhaps also in non-Bayesian directions. Programs seem to be more useful if they possess (7) *explanatory capabilities*, i.e., if they can give an account of how the program has come to a conclusion. MYCIN and several other systems have this capacity; INTERNIST-1 does not, but CADUCEUS does have the possibility for such a capacity. I would also suggest that some (8) *incorporation of utilities* would be desirable for the reasons discussed in the previous section – namely, to rule out dangerous but treatable diseases of low probability. A computer diagnostic program should also be at the same time (9) *comprehensive*, i.e., it should have a reasonably broad scope to obviate the necessity for pre-diagnosis and yet should be (10) *simple*. This desideratum of "simplicity" is intended to eliminate *ad hoc* cumbersomely contructed logics and to place a premium on clarity of the logical basis, consistent, however, with the complexity of the knowledge-base and the reasoning patterns of expert clinicians. Artificial simplicity is certainly worse than an unaesthetic, but diagnostically accurate, complexity. The system should be able to analyze not only cases of single diseases but also (11) *multiple disease diagnoses*. Data from hospital records (and autopsies) indicate that anywhere from 25% to 70% of clinic patients, depending on the clinic, have combinations of diseases ([10], p. 409). Finally, under this heading, a diagnostic system should have the capacity for (12) *temporal development of a disease*.

B. Social Desiderata

Even if logically and empirically adequate solutions to the above list of problems and desiderata for computer diagnosis are forthcoming, there are very likely to exist severe "social" problems. Such difficulties are not the main thrust of this paper but for completeness I believe they ought to be mentioned as additional desiderata. More details on those can be found in Friedman and Gustafson [6] and in Groner *et al.* [9], from which some of these desiderata are taken. Computer diagnosis must be implemented in such a manner that it (13) *interfaces easily with physician and/or patient*. At present, many physicians often find that such programs are novelties that take more of their valuable time that most are willing to offer. In INTERNIST-1, this has been alleviated by development of a frame-based program that prompts the physician in his case entry by presenting successively specific alternatives. Nonetheless, ease of interfacing will not result in widespread acceptance unless it can be demonstrated that computer diagnosis (14) *exceeds the average physician's diagnostic abilities.* MYCIN apparently can do this in the bacteremia area and INTERNIST-1 can occasionally better even the skilled clinician, but more work in the fronts discussed earlier obviously is required. Computer diagnosis will have to (15) be *ultimately cost effective,* (16) be *transferable to new institutions,* (17) *possess adequate privacy safeguards* vis à vis patient's medical records, and (18) *lead to demonstrably better patient care.* Finally, there will have to be (19) *legal clarification for responsibility of computer based diagnostic errors,* though the latter may be moot if the computer diagnostic program is construed as a "tool" rather than a "decision maker".

V. CONCLUSION

In conclusion, there are a plethora of problems in the area of computer diagnosis. I trust, however, that I have indicated that my position is that the problems are likely to be soluble ones and that computer diagnosis is an area that is both conceptually interesting and socially promising.

The George Washington University
Washington, DC, U.S.A.

NOTES

* Grateful acknowledgement is made to the National Endowment for the Humanities for support of my research in philosophy of medicine.
** This essay overlaps in part with an earlier essay of mine [29] and in part with a lengthy essay of mine which serves as an introduction to papers [19], [28], [37] and [38] among others, mentioned at various points in this article. This latter essay is to be found in Schaffner, K.F. (ed.) 1985 *Logic of Discovery and Diagnosis in Medicine,* University of California Press, Berkeley, 1–32, and develops in much more detail than can be accomodated here both the context and further developments of the INTERNIST/CADUCEUS diagnostic program.
[1] The name of the Pittsburgh-based diagnostic program has changed several times since its inception as DIALOGUE, largely for copyright reasons. The stages are DIALOGUE → INTERNIST-1 → INTERNIST-2 → CADUCEUS (see reference [21] for the QMR terminology.) In my discussion I have attempted to use the appropriate name to describe as closely as possible the appropriate stage of development.

BIBLIOGRAPHY

1. Bleich, H.L.: 1972 'Computer-Based Consultation: Electrolyte and Acid-Base Disorders', *American Journal of Medicine* **53** 285 – 291.
2. De Dombal, F.T. and Grémy, F. (eds.): 1976 *Decision Making and Medical Care: Can Information Science Help?* North-Holland, Amsterdam.
3. Feinstein, A.: 1973, 'An Analysis of Diagnostic Reasoning. I The Domains and Disorders of Clinical Macrobiology', *Yale Journal of Biology and Medicine* **46**, 212–232.
4. Feinstein, A.: 1973, 'An Analysis of Diagnostic Reasoning. II The Strategy of Intermediate Decisions', *Yale Journal of Biology and Medicine* **46**, 264–283.
5. Feinstein, A.: 1974, 'An Analysis of Diagnostic Reasoning. III The Construction of Clinical Algorithms', *Yale Journal of Biology and Medicine* **47**, 5–32.
6. Friedman, R.B. and Gustafson, D.H.: 1977, 'Computers in Clinical Medicine, A Critical Review', *Comput. and Biomed. Res.* **10**, 199–204.
7. Gorry, G.A.: 1976, Knowledge-Based Systems for Clinical Problem Solving', in F.T de Dombal and D. Grémy (eds.), *Decision Making and Medical Care: Can Information Science Help?* North-Holland, Amsterdam, pp. 23–31.
8. Gorry, G.A. *et al.*: 1973, 'Decision Analysis as the Basis for Computer-Aided Management of Acute Renal Failure', *American Journal of Medicine* **55**, 473–484.
9. Groner, G.F. *et al.*: 1974, *Applications of Computers in Health Care Delivery: An Overview and Research Agenda*, The Rand Corporation, Santa Monica, California.
10. Habbema, J.D.F.: 1976, 'Models for Diagnosis and Detection of Combinations of Diseases, in F.T. de Dombal and F. Grémy (eds.), *Decision Making and Medical Care: Can Information Science Help?*, North Holland, Amsterdam, pp. 399–410.
11. Harré, R. : 1970, *The Principles of Scientific Thinking,* University of Chicago Press, Chicago, sp. see pp. 157–177
12. Hoffman, P.J.: 1960 'The Paramorphic Representation of Clinical Judgment', *Psychological Bulletin,* **57** 116–131.
13. Jacquez, J.A. (ed.): 1964, *The Diagnostic Process*, Mallory Lithographing, Ann Arbor, Michigan.
14. Jacquez, J.A.(ed.): 1972, *Proceedings of the Second Conference on the Diagnostic Process*, Thomas, Springfield, Illinois.
15. Jacquez, J. and Worusis M.: 1976, The Importance of Symptom Non-Independence in Diagnosis in F.T DeDombal and F. Gremy (eds.), *Decision Making and Medical Care: Can Information Science Help?*, North-Holland, Amsterdam, pp. 379–390.
16. Kleinmuntz, B.: 1965, 'Diagnostic Problem Solving by Computer', *Jap. Psychol. Res.* **7**, 189–194.
17. Ledley, R.S. and Lusted L.B.: 1959, 'Reasoning Foundations of Medical Diagnosis', *Science* **130**, 9–21.
18. Lister, T.A. and J.C. Long : 1978, 'Case Records of the Massachusetts General Hospital *New England Journal of Medicine* **298**, 1133–1138.
19. McMullin, E.: 1985, 'Computer-Aided Diagnosis,' in K.F. Schaffner (ed.), *Logic of Discovery and Diagnosis in Medicine*, University of California Press, Berkeley, pp. 199–222.
20. Miller, R.A., Pople H.E., and Myers, J.D.: 1982, 'INTERNIST-1, An Experimental Computer-Based Diagnostic Consultant for General Internal Medicine,' *New England*

Journal of Medicine **307**, 468–476.

21. Miller, R., Masarie, F.E., and Myers, J.D.: 1986 'Quick Medical Reference (QMR) for Diagnostic Assistance', MD Computing **3**(5), 34–48.
22. Nilsson, N.J.: 1980 *Principles of Artificial Intelligence*, Tioga, Palo Alto.
23. Pauker, S.G. *et al.*: 1976, 'Towards the Simulation of Clinical Cognition', *American Journal of Medicine* **60** 891–996.
24. Pednault, E.P., Zucker, S. W., and Muresan, L.V.: 1981 'On the Independence Assumption Underlying Subjective Bayesian Updating', *Artificial Intelligence* **8**, xxx-xxx.
25. Pipberger, H.V. *et al.*: 1968,' Compiler Evaluation of Statistical Properties of Clinical Information in the Differential Diagnosis of Chest Pain', , *Meth. Information Med.* **7** (1968), 79 – 92.
26. Pople, H.E., Jr.: 1975, 'Dialog [INTERNIST]: A Model of Diagnostic Logic for Internal Medicine', *Advance Papers Int. Joint Conf. Artif. Intell.* **4**, 848–855.
27. Pople, H.E. : 1982, 'Heuristic Methods for Imposing Structure on Ill-Structured Problems: The Structuring of Medical Diagnostics' in Szolovits (ed.), pp. 119–185.
28. Pople, H.E. : 1985, 'Coming to Grips with the Multiple Diagnosis Problem,' in K.F. Schaffner (ed.), *Logic of Discovery and Diagnosis in Medicine*, University of California Press, Berkeley, pp. 181–198.
29. Schaffner, K.F.: 1981, 'Modeling Medical Diagnosis: Logical and Computer Approaches,' *Synthese* **47**, 163–199.
30. Schwartz, W.B. *et al.*: 1973, 'Decision Analysis and Clinical Judgment', *American Journal of Medicine* **55** (1973), 459– 472.
31. Shafer, G.: 1976, *A Mathematical Theory of Evidence*, Princeton University Press, Princeton, New Jersey.
32. Shortliffe, E.H.: 1976, *Computer Based Medical Consultation: MYCIN*, Elsevier, , New York.
33. Shortliffe, E.H. and Buchanan, B.G.: 1975, 'A Model of Inexact Reasoning in Medicine', *Math. Biosciences* 23, 351–379. [Reprinted with some changes in Ch. 4 in shortliffe [27].)
34. Simmons, H.E. and Stolley, P.D.: 1974, 'This is Medical Progress? Trends and Consequences of Antibiotic Use in the United States', *Journal of the American Medical Association* **227**, 1023–1028.
35. Simon, H.: 1977, 'The Structure of Ill-Structured Problems' in his *Models of Discovery*. Reidel, Dordrecht, pp. 304–325. [Originally published in 1973.]
36. Simon, H.: 1981 *Sciences of the Artificial*, M.I.T. Press, Cambridge, MA.
37. Simon, H.: 1985, 'Artificial Intelligence Approaches to Problem Solving and Clinical Diagnosis', in K. Schaffner *et al.* (eds.), *Logic of Discovery and Diagnosis in Medicine*, University of California Press, Berkeley, California, pp. 72–93.
38. Suppe, F.: 1985 'Comments on Pople and Simon', in K.F. Schaffner (ed.), *Logic of Discovery and Diagnosis in Medicine*, University of California Press, Berkeley, pp. 223–242.
39. Szolovits, P. (ed) : 1982, *Artificial Intelligence in Medicine*, Westview Press, Boulder.
40. Szolovitz, P. and Pauker, S.: 1978, Categorical and Probabilistic Reasoning in Medical Diagnosis', *Artificial Intelligence* **II**, 115–144.
41. Törnebohm, H.: 1966 'Two Measures of Evidential Strength', In J. Hintikka and P. Suppes (eds.), *Aspects of Inductive Logic and the Foundations of Mathematics*,

North-Holland Publishing Company, Amsterdam, pp. 81–95.

42. Warner, H.R. *et al.*: 1964, 'Experience with Bayes' Theorem for Computer Diagnosis of Congential Heart Disease', *Ann. N.Y. Acad. Sci.* **115**, 558–567.
43. Whitbeck, C. and Brooks, R.: 1983, 'Criteria for Evaluating a Computer Aid to Clinical Reasoning', *The Journal of Medicine and Philosophy* **8**, 51–65.
44. Winston, P.H.: 1984, *Artificial Intelligence*, 2nd. edit.: Addison-Wesley, Reading Massachusetts.
45. Yu.V. *et al.*: 1979, 'Evaluating the Performance of a Computer Based Consultant', *Computer Programs in Biomedicine* **9**, 95–102.

APPENDIX I

(Reprinted from the *New England Journal of Medicine*, May 18, 1978, with permission.)

CASE RECORDS OF THE MASSACHUSETTS GENERAL HOSPITAL

Weekly Clinicopathological Exercises

ROUNDED BY RICHARD C. CABOT

ROBERT E. SCULLY, M.D., EDITOR
JAMES J. GALDABINI, M.D. ASSOCIATED EDITOR
BETTY U. MCNEELY, ASSISTANT EDITOR

CASE 19–1978

PRESENTATION OF CASE

First admission. A 66-year-old policeman was admitted to the hospital because of nausea.

He was well until one month previously when he began to have occasional night sweats. Ten days before entry he became nauseated, and an uneasy sensation appeared in the upper portion of the abdomen, with rise of the temperature as high as 38.8°C and shaking chills. He lost 0.5 kg in weight during the week preceding admission.

There was a two-year history of mild hypertension that was treated with a mixture of reserpine, hydralazine and hydrochlorothiazide. He stopped his excessive use of cigarettes two years before admission but continued to take one alcoholic drink daily. There was no history of abdominal pain, vomiting, change in bowel habit, hematochezia, melena or dysuria. He had not traveled outside this country.

The temperature was 38.9°C, the pulse 80, and the respirations 18. The blood pressure was 140/70 mm Hg.

On examination the patient appeared well. Abdominal examination disclosed tenderness on percussion over the right costal margin and slight tenderness in the right upper quadrant, where a rounded, smoothmass descended 5 cm; the spleen was not felt.

The urine was normal. The hematocrit was 34.7 per cent the white-cell count was 4500, with 44 per cent neutrophils, 5 per cent band forms, 16 per cent lymphocytes, 28 per cent monocytes, 4 per cent eosinophils and 3 per cent basophils. A stool specimen gave a negative test for occult blood. The urea nitrogen was 19 mg, the bilirubin 0.5 mg, and the protein 7.9 g (the albumin 3.4 g and the globulin 4.5 g) per 100 ml. X-ray films of the chest showed moderate hyperinflation of the lungs, with flattering of the diaphragm; moderate fibrotic changes were visible in both middle-lung fields; the ventricle appeared slightly prominent. On a film of the abdomen the liver and spleen did not appear enlarged A blood culture and a culture of urine yielded no growth of micro-organisms.

On the third hospital day a laparotomy revealed that the liver was enlarged, and a biopsy specimen was obtained; a cholecystectomy was performed. Microscopical examination of the gallbladder was negative. Examination of the specimen of the liver revealed focal infiltration of portal tracts by small lymphocytes, plasma cells and epithelioid-cell granulomas; one granuloma contained scattered large lymphoid cells with vesicular nuclei, single prominent nucleoli and amphophilic cytoplasm; no necrosis of fibrosis was found, and there was no evidence of cholestasis; special stains for micro-organisms were negative. Examination of a lymph node excised from the porta hepatis showed a few small epithelioid-cell granulomas, without atypical lymphoid cells. Postoperatively, the erythrocyte sedimentation rate was 74 mm per hour. A tuberculin skin test (PPD, 5 TU) was negative. A repeated blood culture yielded no growth of micro-organisms. Tests for antimitochondrial and antinuclear antibodies were negative. The patient was discharged on the 12th hospital day.

Second admission (33 days later). In the interval the patient had a daily afternoon temperature rise to 38.3°C, despite frequent doses of aspirin. Drenching night sweats occurred about three times weekly, with nocturia. The hematocrit was 30 per cent, and the erythrocyte sedimentation rate 112 mm per hour. A test for antinuclear antibodies was positive in a titer of 1:16. Tests for rheumatoid factor and cryoprecipitate and a serologic test for syphilis were negative. Serum immunoelectrophoresis demonstrated that considerable normal IgG was present, with an IgG M component identified as the kappa type; there was a mild increase in IgA; IgM was normal. A quantitative determination of the total IgG showed that it was 2720 mg per 100 ml (normal 540 to 1663 mg). Two blood cultures yielded no bacterial growth. Twenty-five days after discharge a segment of the left temporal artery was excised for biopsy; microscopial examination revealed no abnormality. During the week before admission the patient experienced vague, generalized discomfort and nausea. One day before entry the temperature rose to 38.9°C, and he began to vomit tan-colored material. He lost an additional 2 kg in weight between the two admissions.

The temperature was 40°C, the pulse 100, and the respirations 18. The blood pressure was 120/60 mm Hg.

On examination he appeared slightly ill. No jaundice was observed. A firm lymph node, 1 by 3 cm, was felt under the lateral margin of the right pectoralis major muscle, and epitrochlear lymph nodes, 1.5 cm in diameter, were also palpable; there were shotty lymph nodes in the right inguinal region. The edge of the liver extended 4 cm below the right costal margin with a vertical span of 10 cm; the spleen was not felt.

The urine was normal. The hematocrit was 30 per cent; the white-cell count was 4400, with 45 per cent neutrophils , 11 per cent band forms, 20 per cent lymphocytes, 11 per cent atypical lymphocytes, 12 per cent monocytes and 1 per cent eosinophilis. The platelet count was 168,000, and the erythrocyte sedimentation rate 80 mm per hour. A stool specimen gave a negative test for occult blood. The urea nitrogen was 17 mg, the uric acid 3.9 mg, the bilirubin 0.5 mg, the iron 43 μg, the iron-binding capacity 240 μg, and the protein 7.2 g (the albumin 2.6 g, and the globulin 4.6 g) per 100 ml. An x-ray film of the chest was unchanged.

Biopsy specimens of a lymph node and of bone marrow were obtained.

DIFFERENTIAL DIAGNOSIS

DR. T. ANDREW LISTER*. Dr. Ferrucci, may-we review the x-ray films?

DR. JOSEPH T. FERRUCCI, JR.: The films of the abdomen taken on the first admission show no abnormality in the density or architecture of the bones. There is no indication of enlargement of the spleen, although the liver appears somewhat prominent. The films of the chest disclose slight generalized hyperinflation of the lungs with increased translucency, and in the mid-zones there is mild linear interstitial scarring, slightly more marked on the left side, with a suggestion of a honeycomb pattern. A few very small areas of apical pleural thickening are evident bilaterally, with volume loss in both upper lobes. I see no evidence of hilar or mediastinal lymphadenopathy. The left ventricle is slightly enlarged, but the heart is otherwise unremarkable. The appearance in the lungs is that of non-specific interstitial fibrosis, with old granulomatous processes and scarring in the apexes.

DR. LISTER: This 66-year-old man presented initially with fever and mild abdominal symptoms. Over the next two months he remained febrile, and lymphadenopathy developed. Although the differential diagnosis on the first admission was pyrexia of unknown origin, the changes in the physical signs and in the peripheral blood during his illness allows us to narrow the field of possibilities considerably, even before the final biopsies were performed.

Acute bacterial infection was a strong possibility when the patient was first seen. He had a high swinging fever and localizing signs in the abdomen compatible with acute cholecystitis or a gallbladder mass, and I assume that gallbladder disease was the provisional diagnosis that led to the laparotomy on the third hospital day. At that time, however, there were several features that made an acute bacterial infection unlikely. First of all, the patient was fairly well despite the persistent fever. Secondly, there was no increase in neutrophilis, and indeed there was an absolute monocytosis. The latter is uncommon even in cases of typhoid fever, with which one sometimes finds a relative monocytosis. Thirdly, cultures of the blood and urine were negative. The findings at laparotomy clearly ruled out gallbladder disease, and the histologic findings in the liver and lymph-node biopsy specimens made acute bacterial infection very unlikely. The granulomas, as well as the history, are still compatible, however, with other types of infection. Brucellosis can result in a similar histologic picture, but the patient was apparently a city dweller and had not traveled abroad. Histoplasmosis occurs in urban areas and could have produced this clinical picture. Neither of these infections, however, is likely in view of the type of granulomas found and the results of the investigations performed after the laparotomy, and neither is associated with a monoclonal increase of the serum immunoglobulins. Tuberculosis, an obvious possibility in view of the granulomas, even though they were not caseating, is improbable in the face of a negative tuberculin skin test unless the infection was overwhelming which it was not. Syphilis, which causes granulomas and enlargement of the epitrochlear lymph nodes was ruled out by the negative serologic test.

The fever persisted despite aspirin therapy, and by the time of the second admission, about one month after discharge from the hospital, lymphadenopathy had developed. In the interval between the two admissions the erythrocyte sedimentation rate was found to be considerably raised, but tests for antimitochondrial and antinuclear antibodies were negative. Also, a temporal-artery biopsy was negative. With an infectious disease excluded, what are the alternatives? We are told that the patient had received hydralazine for two years before admission. That was certainly long enough to produce a syndrome resembling systemic lupus erythematosus. The clinical features at the time of admission are compatible with that diagnosis, and the test for antinuclear antibodies, which had been negative, became weakly positive. However, after discharge the illness progressed, and such a sequence is infrequent in cases of this syndrome. Other factors making this diagnosis unlikely are the monoclonal increase in the serum immunoglobulins, as opposed to the polyclonal increase that is usually seen with lupus, and the monocytosis followed by the appearance of abnormal lymphocytes in the peripheral blood.

Sarcoidosis is obviously a consideration because of the fever, the lymphadenopathy, particularly involving epitrochlear nodes, and the non-caseating granulomas; the absence of a rash, iritis or any of the other features of the disease does not exclude it. The laboratory findings in cases of sarcoidosis are usually not very specific. The negative tuberculin skin test is the only finding to suggest the diagnosis. Although some observers consider monocytosis a feature of the disease,[1] it is in fact rare. Immunoglobulin changes do occur, but as in cases of lupus they are almost always polyclonal. It is also possible that the granulomatous process in the liver was the entity described by Simon and Wolf, [2] who discussed 13 patients with granulomatous hepatitis and prolonged fever of undetermined origin, but lymphadenopathy was not reported in that group of patients.

One possible diagnosis that should be mentioned is angioimmunoblastic lymphadenopathy.[3,4] Patients with his disorder are generally over 50 years of age. It is characterized by fever, generalized lymphadenopathy, hepatomegaly, splenomegaly and frequently a maculopapular skin rash. When initially seen the patients are often not very ill despite the farily striking physical findings. Therefore, the presentation of this patient fits with that diagnosis, but the laboratory findings are not consistent with it. Although a monocytosis and abnormal lymphocytes may be seen in the blood, the immunoglobulin changes are almost always polyclonal. More importantly, discrete, non-caseous granulomas in the liver would not be expected to occur in association with this disease.

I have used the presence of a monoclonal increase in the serum IgG immunoglobulin to rule out several of the preceding diagnoses. This conclusion seems justified in the case under discussion, but it is important to remember that although benign M components may be of the IgG class they more often contain IgM heavy chains. Ten per cent of a population of subjects about the age of this man who were studied in Sweden[5] were found to have monoclonal immunoglobulinemia. During a follow-up period of 2½ years myeloma developed in only two of them. The authors commented that the serum M component in this group of subjects was usually less than 1 g per 100 ml.

Thus far, I have not mentioned the possibility of a metastatic carcinoma from an occult source. The distribution of the lymphadenopathy in this case makes that diagnosis unlikely. Also, the granulomas that occur in association with carcinoma usually appear only in lymph nodes draining the tumor.

A malignant lymphoma associated with or obscured by granulomas is a much more likely possibility. The use of laparotomy for the staging of Hodgkin's disease and other lymphomas has provided the opportunity for examination of lymph nodes, the liver and the spleen. Recent studies from the Stanford Medical Center[6-8] have demonstrated that non-caseating granulomas may occur in these sites with or without coexistent lymphoma evident within the abdomen. The confusion that can be caused by such granulomas, particularly when they occur in association with non-Hodgkin lymphoma, has been emphasized in a recent report,[9] and their meaning in patients with lymphoma has been disputed.

My diagnosis is a malignant lymphoma, probably of the diffuse lymphocytic type, associated with non-caseating granulomas in the liver. This diagnosis is compatible with all the clinical findings and most of the laboratory data. Monocytosis occurs frequently with non-Hodgkin lymphoma, and so, of course, do abnormal lymphocytes in the peripheral blood. The monoclonal immunoglobulinemia is also explained, as is the negative tuberculin skin test. In making this diagnosis I am obliged to ignore the positive test for antinuclear antibodies, which is almost never observed in a patient with Hodgkin's disease or any other type of lymphoma. However, the titer was so low that I shall disregard that finding. The only other deterrent to the diagnosis of non-Hodgkin lymphoma is the appearance of the granulomas in the liver-biopsy specimen, with the possibility that Reed-Sternberg cells were present in them. If they were, and if the patient did have Hodgkin's disease, I would have to conclude that he had a concomitant benign monoclonal gammopathy. I suspect that the final lymphnode biopsy revealed a diffuse lymphocytic lymphoma, and that the bone-marrow biopsy showed either diffuse or focal infiltration.

DR. ROBERT E. SCULLY: Dr. Jacobson, what diagnosis would you make in this case?

DR. BERNARD M. JACOBSON: The most common cause of an IgG monoclonal M component is multiple myeloma, and next in order of frequency are various types of lymphoma, including the Hodgkin and non-Hodgkin types. For that reason my diagnosis is also some form of lymphoma.

DR. SCULLY: Dr. Aisenberg, what is your opinion?

DR. ALAN C. AISENBERG: I find it difficult to choose between Hodgkin's disease and non-Hodgkin lymphoma. I am unwilling to use the IgG M component as a deciding point because it could have been benign and unrelated, as Dr. Lister mentioned, particularly in view of the patient's age. The M-component level, which was between 1 and 2 g per 100 ml of serum, and the considerable amount of normal IgG immunoglobulin favor a benign monoclonal gammopathy. If the M component had been of an IgM class it would be inconsistent with Hodgkin's disease.

The febrile course is much more consistent with Hodgkin's disease or possibly a diffuse histiocytic lymphoma than other forms of lymphoma, miliary granulomas are more common with Hodgkin's disease than non-Hodgkin lymphoma, and the clinical course is also a little more in favor of the former. On the other hand, I have never seen bilateral involvement of the epitrochlear lymph nodes in a patient with Hodgkin's disease, although I have seen a few cases with unilateral involvement. Bilateral involvement is more frequent in patients with non-Hodgkin lymphoma, and on that very weak clinical basis I would pick a non-Hodgkin lymphoma.

DR. JACOBSON: I do not agree that the M component in this patient reflected a benign reflected a benign gammopathy. It is very uncommon in this hospital to find an IgM or IgG M component that is not associated with severe disease. Moreover, the frequency of benign M components is low. Of 47 subjects between the ages of 50 and 59 years studies by Kyle and his associates[10] the prevalence of benign monoclonal immunoglobulinemia was only 0.5 per cent. Among 147 subjects 80 years of age or older it was 4.8 per cent.

DR. LEONARD ELLMAN: It is worth noting that it may make a difference whether one is surveying patients in a hospital or healthy persons. A much greater importance may have to be attached to an M component that is detected in a patient in the hospital. However, the relatively high frequency of benign M components in elderly persons is widely accepted now.

DR. SCULLY: Dr. Schoolnik, will you give us the medical student's diagnoses?

DR. GARY K. SCHOOLNIK: They considered idiopathic granulomatous hepatitis unlikely because of the presence of peripheral lympahdenopathy. The IgG M component was interpreted as either incidental and related to the patient's age or a reflection of a lymphoreticular malignant tumor. Their differential diagnosis included sarcoidosis, malignant lymphoma and disseminated tuberculosis.

CLINICAL DIAGNOSIS

Malignant lymphoma.

DR. T. ANDREW LISTER'S DIAGNOSIS

Malignant lymphoma, associated with noncaseating granulomas,? type.

PATHOLOGICAL DISCUSSION

DR. JOHN C. LONG: Examination of a specimen of bone marrow obtained six weeks after the liver biopsy revealed partial replacement by an infiltrate of lymphocytes, plasma cells, eosinophils and epithelioid histiocytes. Discrete granulomas were not present, and there was no necrosis. A silver impregnation stain disclosed a slight increase of the reticulin fibers in the involved area. There were large lymphoid cells with prominent nucleoli, and a single diagnostic Reed-Sternberg cell was identified. A smear of the marrow aspirate showed a slight increase in eosinophils and plasma cells but was otherwise negative.

An axillary-lymph-node biopsy done four days later revealed extensive granulomatous replacement. Numerous epithelioid cells were disposed in confluent sheets and cohesive clusters, with many multinucleated giant cells of the Langhans type. Interspersed among the histiocytes were small lymphocytes, plasma cells, eosinophils and rare Reed-Sternberg cells. Vascular involvement was prominent, with a granulomatous cellular infiltrate of the walls of small arteries within the perinodal fat. Immunoperoxidase stains of the lymph-node-biopsy specimen, done by Dr. Masamichi Kojiro, of our laboratory, revealed the presence of lysozyme within many epithelioid cells in the granulomas.[11] This case emphasizes the difficulty that may be encountered by the pathologist in the diagnosis of Hodgkin's disease. The exuberant granulomatous component obscured the underlying lymphoid infiltrate in the axillary-lymph-node specimen. The abnormality in the liver-biopsy specimen was originally interpreted as granulomatous hepatitis, consistent with sarcoidosis or primary biliary cirrhosis. Givler and his associates[12] have empahsized that the hepatic infiltrates of Hodgkin's disease may vary in cellular composition. The involvement may stimulate benign granulomatous disease, as in the present case, or can be pleomorphic and fibrotic with a paucity of atypical cells. We detected the large, abnormal cells within the hepatic granulomas only after the axillary-lymph-node and bone-marrow biopsy specimens had been examined. The findings were then interpreted as Hodgkin's disease of the mixed-cellularity type with involvement of lymph nodes, bone marrow and liver. The diagnostic Reed-Sternberg cell in the bone marrow provided the first clue that we were dealing with Hodgkin's disease rather than a benign, disseminated granulomatous disorder. At least three patterns of bone-marrow involvement may be observed in Hodgkin's disease.[13] The first consists of focal granulomas, which may exhibit minimal cytologic atypicality and may be erroenously interpreted as infectious granulomas.[14] The second pattern is that of diffuse fibrosis, which may simulate idiopathic myelofibrosis. Finally, there may be discrete tumor nodules with abnormal cells, reticular fibrosis and characteristic Reed-Sternberg cells. Silver impregnation stains to demonstrate abnormal connective tissue with a focal deposition of disorderly reticulin fibers may be helpful in the evaluation of bone marrow suspected of being involved with Hodgkin's disease.

DR. SCULLY: Should the presence of the atypical lymphoid cells in the hepatic granulomas have been more suggestive of a Hodgkin than a non-Hodgkin lymphoma at the time of the liver biopsy?

DR. LONG: Yes. The presence of mononuclear Reed-Sternberg variants in the liver-biopsy specimen is more compatible with Hodgkin's disease than a lymphocytic lymphoma with epithelioid-cell granulomas.

DR. AISENBERG: This process doesn't fit into the spectrum of Lennert's lesion does it?

DR. LONG: No. The Lennert lesion[15] is characterized by numerous epithelioid cells disposed diffusely or in clusters reminiscent of the pattern observed in toxoplasma lymphadenitis, in contrast to the well formed, discrete granulomas with multinucleated giant cells seen in this case. Moreover, in Lennert's lesion the epithelioid cells are admixed with reticular lymphoblasts, plasmacytoid cells and lymphocytes with irregular clefted nuclei.[16–17]

DR. LISTER: Dr. Aisenbery, how often do you see abnormal lymphocytes in the peripheral blood of patients with Hodgkin's disease?

DR. AISENBERG: A florid picture with many atypical lymphocytes containing cleaved nuclei is not seen in Hodgkin's disease. I have not seen the smear in the present case, but the findings as described do not sound diagnostic of non-Hodgkin lymphoma. The presence of a few atypical lymphocytes is not unusual in patients with Hodgkin's disease, and the picture in this case was balanced by a monocytosis. All in all, I would have said that the picture observed in the smear is probably compatible with Hodgkin's disease.

DR. ELLMAN: It is generally believed that older patients with Stage IVB Hodgkin's disease have a poor prognosis, and this patient's illness was considered very serious. Also, he had neutropenia, which compromised his ability to receive intensive chemotherapy. After some thought I elected to put him on a protocol that involves the cyclical use of lomustine (CCNU), vinblastine, procarbazine and prednisone. After two courses of his treatment a fairly severe granulocytopenia developed, but otherwise there have been no serious toxic effects. He appears to be having a very favorable response. The systemic symptoms and the lymphadenopathy have disappeared entirely.

DR. LONG: A repeated serum protein immunoelectrophoresis done six months after the previous study revealed a mild, diffuse increase in the immunoglobulins without a detectable M component.

ANATOMICAL DIAGNOSIS

Hodgkin's disease, mixed-cellularity type.

NOTE

* Visiting fellow in medical oncology, Sidney Farber Cancer Institute senior lecturer and honorary consultant physician. Department of Medical Oncology, St. Bartholomew's Hospital, London, England.

REFERENCES

1. Scadding JG: Sarcoidosis. London, Eyre & Spottiswoode, 1967
2. Simon HB. Wolff SM: Granulomatous hepatitis and prolonged fever of unknown origin: a study of 13 patients. Medicine (Baltimore) 52:1–21 1973.
3. Lukes RJ, Tindle BH: Immunoblastic lymphadenopathy: a hyperimmune entity resembling Hodgkin's disease N Engl J Med 292:1–8, 1975
4. Frizzera G, Moran EM, Rappaport H: Angio-immunoblastie lymphadenopathy: diagnosis and clinical course. Am J Med 59: 803–818, 1975.
5. Axelson U, Hällen J: Review of fifty-four subjects with monoclonal gammopathy. Br J Haematol 15: 417–120, 1968
6. Kadin ME, Donaldson SS. Dorfman RF: Isolated granulomas in Hodgkin' disease N Engl J Med 283: 859–861, 1970
7. Kadin ME, Glatstein E, Dorfman RF: Clinicopathologic studies of 117 untreated patients subjected to laparotomy for the staging of Hodgkin's disease. Cancer 27: 1277–1294, 1971
8. Kim H. Dorfman RF: Morphological studies of 84 untreated patients subjected to laparotomy for the staging of non-Hodgkin's lymphomas. Cancer 33:657–674, 1974
9. Braylan RC, Long JC, Jaffe ES, et al: Malignant lymphomas obscured by concomitant extensive epithelioid granulomas. Cancer 39: 1146–1155, 1977
10. Kyle RA, Finkelstein S, Elveback LR, et al: Incidence of monocional proteins in a Minnesota community with a cluster of multiple myeloma. Blood 40: 719–724, 1972
11. Pinkus GS, Said JW: Profile of intracytoplasmic lysozyme in normal tissues, myeloproliferative disorders, hairy cell leukemia and other pathologic processes: an immunoperoxidase study of paraffin sections and smears Am J Pathol 89: 351–366, 1977
12. Givler RL, Brunk SF, Hass CA, et al: Problems of interpretation of liver biopsy in Hodgkin's disease. Cancer 28: 1335–1342, 1971
13. O'Carroll DI. McKenna RW, Brunning RD: Bone marrow manifestations of Hodgkin's disease. Cancer 38: 1717–1728. 1976
14. Case Records of the Massachusetts General Hospital (Case 11–1973). N Engl J Med 288: 570–576, 1973
15. Lennert K, Mestdagh J: Lymphocranulomatosen mit konstant hohem Epitheloidrellgehalt. Virchowa Arch [Pathol Anat] 344: 1–20, 1968
16. Burke JS. Butler JJ: Malignant lymphoma with a high content of epithelioid histiocytes (Lennert's hyphoma). Am J Clin Pathol 66:1–9, 1976
17. Case Records of the Massachusetts General Hospital (Case 30–1977). N Engl J Med 297: 206–211, 1977

APPENDIX II

(DOCTOR)
ENTER CASE NAME: NEJMV298P1133
INITIAL POSITIVE MANIFESTATIONS:
AGE GTS THAN 55
SEX MALE
SKIN SWEATING INCREASED GENERALIZED
FEVER
RIGOR <S>
HYPERTENSION HX
CIGARETTE SMOKING HX
ABDOMEN TENDERNESS RIGHT UPPER QUADRANT
ABDOMEN MASS RIGHT UPPER QUADRANT
HEMATOCIT BLOOD LESS THAN 35
ELECTORPHORESIS SERUM GAMMA GLOBULIN INCREASED
CHEST XRAY LUNG <S> LUCENCY INCREASED GENERALIZED
CHEST XRAY DIAPHRAGM LOW BILATERAL
CHEST XRAY INTERSTITIAL MARKINGS INCREASED
HEART XRAY LEFT VENTRICLE ENLARGED
TUBERCULIN SKIN TEST NEGATIVE
LIVER BIOPSY GRANULOMA <S> NON CASEATING
LYMPH NODE BIOPSY GRANULOMA <S> NON CASEATING
ANA POSITIVE
IMMUNOELECTROPHORESIS SERUM IGG INCREASED
ELECTROPHORESIS SERUM GLOBULIN SPIKE
VOMITING RECENT
LYMPH NODE <S> ENLARGED
LIVER ENLARGED SLIGHT
PLATELETS 50,000 TO 20,000
ELECTROPHORESIS SERUM ALBUMIN DECREASED
IRON SERUM LESS THAN 45

INITIAL NEGATIVE FINDINGS:
DIARRHEA ACUTE
CONSTIPATION
ABDOMEN PAIN EPIGASTRIUM
ABDOMEN PAIN RIGHT UPPER QUADRANT
FECES GROSS BLOOD
FECES BLACK TARRY
DYSURIA
RESIDENCE OR TRAVEL TROPICAL OR SEMITROPICAL HX
TACHYCARDIA
TACHYPNEA
SPLENOMEGALY SLIGHT
SPLENOMEGALY MODERATE
PROTENURIA

WBC 4,000 TO 14,000 PERCENT NEUTROPHIL <S> INCREASED
WBC 14,000 TO 30,000
FECES GUAIAC TEST POSITIVE
UREA NITROGEN BLOOD 30 TO 60
UREA NITROGEN BLOOK GTR THAN 60
BILIRUBIN BLOOD CONJUGATED INCREASED
URINE CULTURE SIGNIFICANT E COLI
ANITBODY MITOCHONDRIAL
CHEST XRAY HILAR ADENOPATHY UNILATERAL
CHEST XRAY HILAR ADENOPATHY BILATERAL

CHEST XRAY MEDIASTINUM WIDE
LIVER BIOPSY FOCAL NECROSIS AND INFLAMMATION
LIVER BIOPSY PERIPORTAL FIBROSIS MILD
LIVER BIOPSY PERIPORTAL FIBROSIS MARKED
LIVER BIOPSY BILE PLUGGING
RHEUMATOID FACTOR POSITIVE
CRYOGLOBULIN <S> BLOOD INCREASED
VDRL POSITIVE
IMMUNOELECTROPHORESIS SERUM IGM INCREASED
ARTERY <IES> CRANIAL SUPERFICIAL BIOPSY ARTERITIS
JAUNDICE
IRON BINDING CAPACITY TOTAL GTR THN 370
IRON BUILDING CAPACITY TOTAL LESS THAN 240

GC: 8
8068, 20356 FREE WORDS

GC: 8
7812, 20100 FREE WORDS

GC: 8
7546, 20346 FREE WORDS

DISREGARDING: ABDOMEN MASS RIGHT UPPER QUADRANT, ABDOMEN TENDERNESS RIGHT UPPER QUADRANT, CHEST XRAY DIAPHRAGM LOW BILATERAL, CHEST XRAY INTERSTITIAL MARKINGS INCREASED, CHEST XRAY LUNG <S> LUCENCY INCREASED GENERALIZED, HEART XRAY LEFT VENTRICLE ENLARGED, ANA POSITIVE, IRON SERUM LESS THAN 45

CONSIDERING: AGE GTR THAN 55, SEX MALE, VOMITING RECENT, FEVER, LIVER ENLARGED SLIGHT, LYMPH NODE <S> ENLARGED, RIGOR <S> SKIN SWEATING INCREASED GENERALIZED, HEMATOCRIT BLOOD LESS THAN 35, ELECTROPHORESIS SERUM ALBUMIN DECREASED, ELECTROPHORESIS SERUM GAMMA GLOBULIN INCREASED, ELECTROPHORESIS SERUM GLOBU-LIN SPIKE, IMMUNOELECTROPHORESIS SERUM IGG INCREASED, PLATELETS 50,000 TO 200,000, TUBERCULIN SKIN TEST NEGATIVE, LIVER BIOPSY GRAN-

ULOMA <S> NOW CASEATING, LYMPH NODE BIOPSY GRANULOMA <S> NON CASEATING

RULEOUT: SYSTEMIC HODGKINS DISEASE, DISSEMINATED ATYPICAL MYCOBACTERIAL INFECTION, DISSEMINATED HISTOPLALSMOSIS, DISSEMINATED TUBERCULOSIS, CHRONIC SYSTEMIC SARCOIDOSIS

Please Enter Findings of APPETITE/WEIGHT
*YIELD
##

*HYDRALAZINE ADMINISTRATION HX

*LIVER BIOPSY PERIPORTAL R##

*LIVER BIOPSY PERIPORTAL FIBROSIS XXX
##

*LIVER BIOPSY ROUN##

*LIVER BIOPSY PERIPORTAL

       [ADMISSIBLE NEXT WORDS: FIBROSIS, INFILTRATION]
...INFILTRATION
       [ADMISSIBLE NEXT WORDS: NEUTROPHIL, ROUND]
...ROUND CELL<S>
*WBC MONOCYTE <S> INCREASED

*GO

GC: 8
9324. 20076 FREE WORDS, 21 PAGES LEFT

DISREGARDING: ABDOMEN MASS RIGHT UPPER QUADRANT, ABDOMEN TENDERNESS RIGHT UPPER QUADRANT, CHEST XRAY DIAPHRAGM LOW BILATERAL, CHEST XRAY INTERSTITIAL MARKINGS INCREASED, CHEST XRAY LUNG <S> LUCENCY INCREASED GENERALIZED, HEART XRAY LEFT VENTRICLE ENLARGED, ANA POSITIVE, IRON SERUM LESS THAN 45, LIVER BIOPSY PERIPORTAL INFILTRATION ROUND CELL <S>

CONSIDERING: AGE GTR THAN 55, SEX MALE, VOMITING RECENT, FEVER, LIVER ENLARGED SLIGHT, LYMPH NODE <S> ENLARGED, RIGOR <S> SKIN SWEATING INCREASED GENERALIZED, HEMATOCRIT BLOOD LESS THAN 35, WBC MONOCYTE <S> INCREASED ELECTROPHORESIS SERUM ALBUMIN DECREASED, ELECTROPHORESIS SERUM GAMMA GLOBULIN INCREASED, ELECTROPHORESIS SERUM GLOBULIN SPIKE, IMMUNOELECTROPHORESIS SERUM IGG INCREASED, PLATELETS 50,000 TO 200,000, TUBERCULIN SKIN

TEST NEGATIVE, LIVER BIOPSY GRANULOMA <S> NON CASEATING, LYMPH NODE BIOPSY GRANULOMA <S> NON CASEATING

PURSUING: SYSTEMATIC HODGKINS DISEASE

Please Enter Findings of Lymph Node <S> Exam

*GO

LYMPH NODE <S> PAIN WITH ALCOHOL?
N/A

Please Enter Findings of BLOOD COUNT AND SMEAR
*GO

WBC REED STERNBERG CELL <S>?
NO

DISREGARDING: ABDOMEN MASS RIGHT UPPER QUADRANT, ABDOMEN TENDERNESS RIGHT UPPER QUADRANT, CHEST XRAY DIAPHRAGM LOW BILATERAL, CHEST XRAY INTERSTITIAL MARKINGS INCREASED, CHEST XRAY LUNG<S> LUCENCY INCREASED GENERALIZED, HEART XRAY LEFT VENTRICLE ENLARGED, ANA POSITIVE, IRON SERUM LESS THAN 45, LIVER BIOPSY PERIPORTAL INFILTRATION ROUND CELL <S>

CONSIDERING: AGE GTR THAN 55, SEX MALE, VOMITING RECENT, FEVER, LIVER ENLARGED SLIGHT, LYMPH NODE <S> ENLARGED, RIGOR <S> SKIN SWEATING INCREASED GENERALIZED, HEMATOCRIT BLOOD LESS THAN 35, WBC MONOCYTE <S> INCREASED, ELECTROPHORESIS SERUM ALBUMIN DECREASED, ELECTROPHORESIS SERUM GAMMA GLOBULIN INCREASED, ELECTROPHORESIS SERUM IGG INCREASED, PLATELETS 50,000 TO 200,000, TUBERCULIN SKIN TEST NEGATIVE, LIVER BIOPSY GRANULOMA <S> NON CASEATING, LYMPH NODE BIOPSY GRANULOMA <S> NON CASEATING

PURSUING: SYSTEMIC HODGKINS DISEASE

Please Enter Findings of BONE MARROW BIOPSY
*GO

BONE MARROW BIOPSY REED STERNBERG CELL <S>?
N/A

Please Enter Findings of SKIN BIOPSY
*OM/M//O/GO

SKIN BIOPSY REED STERNBERG CELL <S>?
N/A

Please Enter Findings of LYMPH NODE BIOPSY
*GO

LYMPH NODE BIOPSY REED STERNBERG CELLS <S>?
N/A

LYMPH NODE BIOPSY EOSINOPHIL <S> INCREASED?
N/A

Please Enter Findings of LYMPHANGIOGRAPHY
*OMIT

CONCLUDE: SYSTEMIC HODGKINS DISEASE

DISREGARDING: CHEST XRAY DIAPHRAGM LOW BILATERAL, CHEST XRAY INTERSTITIAL MARKINGS INCREASED, CHEST XRAY LUNG <S> LUCENCY INCREASED GENERALIZED, HEART XRAY LEFT VENTRICLE ENLARGED, ANA POSITIVE, IRON SERUM LESS THAN 45, LIVER BIOPSY PERIPORTAL INFIL-TRATION ROUND CELL <S>

CONSIDERING: ABDOMEN MASS RIGHT UPPER QUADRANT, ABDOMEN TEN-DERNESS RIGHT UPPER QUANDRANT

RULEOUT: LYMPHOMA OF COLON, GASTRIC LYMPHOMA, HEPATIC HEMAN-GIOMA, PERITONEAL NEOPLASM, CYSTIC DISEASE OF LIVER, SMALL INTESTINAL NEOPLASM
Please Enter Findings of PAIN ABDOMEN
*

....

HENRIK R. WULFF

COMPUTERS AND CLINICAL THINKING

Kenneth H. Schaffner [14] gives a fair account of the state of computer diagnosis and Edmund D. Pellegrino [13] discusses, as I shall do, the clinical implications of accepting the machine as a partner. I enjoyed reading both papers and do not disagree with Pellegrino's points of view, but I shall approach the problems from a somewhat different angle.

I am sure that Pellegrino is right when he prophesies that "the machine is inescapably part of the future of medicine", but it is worth stressing that the partnership between clinician and computer will never be an equal one. The clinician will never be able to share the responsibility for his decisions with the computer, and perhaps the fear of a dehumanizing effect of computer technology could be allayed by substituting *computer-aided* diagnosis for the term computer diagnosis. However, clinical decisions made with the aid of a computer will of course reflect the way in which the computer "thinks", and we must consider the possibility that in the future clinicians' diagnoses will be based to a considerable extent on statistical decision theory and computer logic. In this essay, I shall discuss the philosophical basis of that particular "mode of thought" and its relationship with conventional clinical thinking.

Until now no one has managed to analyse in full detail the clinical decision process, and the main difficulty may well be that clinical reasoning is a synthesis of three different ways of thinking that may be called *biological, empirical*, and *humanistic*. I shall first explain these terms and then return to the topic of computer-aided diagnosis.

Biological thinking gained access to bedside medicine in the last century, when doctors began to study systematically the structure and functions of the human organism in health and disease. Since then medical scientists have seen it as their goal to acquire as much biological knowledge as is possible in order to enable clinicians to deduce from biological theory what is wrong in the individual case and which is the best treatment. If, for instance, a thyroidectomized patient develops tetany, the clinician will deduce from his theoretical knowledge that the tetany is caused by hypocalcaemia and that the hypocalcaemia is caused by accidental removal of the parathyroids during the surgical operation.

J. L. Peset and D. Gracia (eds.), The Ethics of Diagnosis, 243–254.

He will confirm the diagnosis by measuring the serum calcium. He then will deduce from his biological knowledge that the patient must be treated with intravenous calcium and later with vitamin-D preparations.

Clinical decisions are not only based, however, on deductions from biological knowledge. The clinician also makes use of *empirical thinking*, which means that he also bases his decisions on uncontrolled and controlled clinical experience. The clinician treats pneumococcal infections with penicillin, because he knows that pneumococci are sensitive to this antibiotic (biological thinking), but also because it has been experienced again and again that penicillin treatment usually cures patients with pneumococcal infections (uncontrolled empirical thinking). But even the combination of uncontrolled experience and biological thinking does not provide a sufficiently reliable basis of clinical decisions. First, clinicians often overestimate the reliability of uncontrolled experience. There are numerous examples in the history of medicine that doctors firmly believed in the efficiency of treatments that later proved quite ineffective or even harmful, e.g., blood-letting in acute infections and anticoagulant treatment of acute myocardial infarction. Second, they also tend to over-estimate the extent of their biological knowledge and the reliability of their deductions. Doctors believed, for instance, that it was self-evident from a biological point of view that premature babies with respiratory distress needed intensive oxygen treatment but, later, it was realized that this treatment caused retrolenticular fibrosis.

Examples of this kind have taught the medical profession that biological and uncontrolled empirical thinking must be supplemented with controlled experience in the form of the results of properly conducted clinical research. It has become generally accepted, especially during the last two decades, that it is necessary to test the effect of new treatments by means of randomized trials and it is also realized that, whenever possible, the reliability of diagnostic methods must be assessed by determining their diagnostic specificity and sensitivity [16]. Studies of that kind require the use of bio-statistical methods, and today readers of medical journals are expected to be acquainted with the basic principle of statistical theory.

Both biological and (controlled or uncontrolled) empirical thinking belong to the realm of natural science: biological theory may tell the clinician how to interfere with the disease mechanism and a randomized controlled trial may help him to choose the best, i.e., the most effective, treatment for that purpose, but, as Pellegrino rightly points out, the goodness of a clinical decision is not just a question of efficacy: a *good* clinical decision is one that is in the *best* interests of this patient. The words "good" and "best" in that context have their full ethical meaning, as they do not refer to that which is good for a particular purpose, but that which is good in itself. In this way, all clinical decision-making transgresses natural science, and the clinician must engage himself in that which I have called *humanistic* thinking. He must remember that the patient is not only a biological organism, but also a self-reflecting human being who reasons, plans ahead, and acts in accordance with his self-chosen values ([17], pp. 121–34). Therefore, it is not sufficient to *explain* the causes of the patient's illness in biological terms and to treat the patient accordingly; it is necessary to *understand* in which way the patient relates himself to his illness and to assign the patient an active role in the decision process.

How, then, does computer diagnosis fit into this triad of thinking?

In his paper, Schaffner discusses branching logic (see fig. 1, p. 203). Flow charts of this kind may, of course, take into account any type of reasoning (apart from concern for the wishes of the individual patient), but it is no worse and no better than the judgment of the clinician who designed the program. In principle, the use of computers for this purpose differs little from asking a colleague for his advice, and such programs are quite harmless, if only the user appreciates that mediation of the advice by a computer does not make it more reliable. Branching logic illustrates that it is not so much the introduction of computer technology that matters. Rather, what is important is the type of program that is fed into the computer.

From a theoretical point of view, the introduction of Bayesian probabilistic diagnosis[1] is much more interesting. It is this type of computer diagnosis that to date has yielded the most promising results. As Schaffner explains, the procedure is simple. The research worker collects a large number of patients presenting a particular problem (i.e., an acute abdomen or jaundice), records the relevant symptoms, signs, and laboratory data in each case, establishes the true diagnosis of each patient, and feeds all this information into a computer. Then, other clinicians, who have recorded the presence and absence (or magnitude) of the same symptoms, signs, and laboratory data in their patients, can use this data base to calculate the probability of different diseases. The conversion of the conditional probabilities is done by means of Bayes' theorem and Schaffner discusses some of the inherent difficulties of this approach (the independence assumption and "the combinatorial problem of nightmarish proportions", if all sub-types of disease are to be catered for). Nevertheless, the procedure is often quite successful in practice. Bayesian probabilistic diagnosis may be regarded as formalized "empirical thinking" and it has been shown repeatedly that computers diagnosing the causes of an acute abdomen or jaundice compete well with skilled clinicians (e.g., [5]. [12]). The Bayesian probabilistic approach has also proved extremely useful for predicting the prognosis of patients who have suffered a severe head injury [7].

It is possible that computer diagnosis by means of Bayes' Theorem will be used more extensively in the future for the solution of well-defined clinical problems, but the spin-offs of the development of such programs may be even more important. De Dombal, pioneer in this field, has pointed out that medical data are much too "fuzzy". He stresses the importance of a *standard, pre-defined terminology amongst doctors* and his group has succeded in defining a terminology for acute abdominal pain that lowered the inter-observer variation by about four-fifths (from 20% to 4%) [4]. It has been shown by numerous studies, especially after the introduction of kappa-statistics [10], that the inter-observer variation in routine clinical practice is appalling, and it may be hoped that the introduction of computer-aided diagnosis may help to open the eyes of clinicians to this problem. Clinical decision-making is also hampered by the fact that clinicians do not always know which questions to ask and which investigations to do, but the experience from computer diagnosis may help to solve this problem. A carefully collected data base may tell us which symptoms and signs have the greatest discriminatory value, and it may also reveal in which sequence questions should be asked and examinations done in order to reach a diagnosis as quickly and economically as possible.

However, it is also important to stress the limitations of Bayesian probabilistic diagnosis. In many cases the structure of the disease classification forbids the use of the probabilistic approach, and in order to explain this important point (which is often overlooked by proponents of computer diagnosis), I shall give the following three clinical examples:

(a) A surgeon states that there is a 90% probability that a patient suffers from acute appendicitis. This statement may well be the result of empirical thinking, as it is possible that the surgeon has studied a large number of similar cases and that he has experienced that 90% of such patients at operation were found to suffer from that disease.

(b) A cardiologist states that there is a 90% probability that a patient suffers from acute myocardial infarction. This statement is not the result of empirical thinking as most patients suspected of this disease survive, and in all such cases the truth or falsehood of the diagnosis cannot be established with certainty. The cardiologist cannot possibly know from experience that 90% of "similar" patients suffer from acute myocardial infarction, but he can deduce from his theoretical knowledge of the significance of, for example, electrocardio-graphic changes and enzyme changes that this particular patient *probably* suffers from that disease. The probability in this case is a subjective probability (a measure of belief), not a frequential one based on past experience.

(c) A rheumatologist states that there is a 90% probability that a patient suffers from systemic lupus erythematosus (SLE). In this case, the probability statement is based neither on experience nor on biological knowledge; it simply does not make very much sense as SLE is a clinical syndrome, the diagnosis of which is based on a global assessment of the observed clinical picture. Since there is no generally accepted definition of this syndrome, it is almost a matter of a opinion whether or not a particular patient is said to suffer from the disease.

In other words, the clinician who has suggested a diagnosis may find himself in one of three situations: (a) The diagnosis is right or wrong, and it is possible to establish the truth by independent means, (b) the diagnosis is right or wrong, but it is not possible to establish the truth with certainty in the majority of cases, and (c) the disease entity is a clinical syndrome and there is no "diagnostic truth" beyond the recorded symptoms and signs [16].

Bayesian probabilistic diagnosis, which presupposes that it is possible to establish the true diagnosis in all those patients who constitute the data base, may be quite useful in the first of these situations, but it is useless in the other two. Empirical diagnostic thinking, which is based on the recorded frequency of different diseases and the recorded frequency of symptoms and signs in those diseases, only serves to solve a minority of those diagnostic problems which clinicians face in the daily routine. This problem has no easy solution as it reflects the very nature of the current disease classification, but the implications would be less serious, if clinicians agreed on standardized working definitions of clinical syndromes (such as SLE) and of those diseases (such as myocardial infarction), which in many cases cannot be diagnosed with certainty.

Pellegrino rightly points out that, in clinical practice, "diagnostic categories have a totally instrumental objective", and for that reason it is not possible to distinguish sharply between the diagnostic decision, the therapeutic decision, and prognostic considerations. This point of view is well illustrated by modern statistical decision theory that combines branching logic, probabilistic considerations, and so-called utility calculations. By such means, attempts have been made to calculate which is the best decision in concrete clinical situations and this approach to clinical decision-making, which in practice also requires the use of computers, is interesting from both a medical and a philosophical point of view.

The procedure in its simplest form is the following: the possible decisions are listed, the probability of different outcomes of these decisions is stated, and the utility of these outcomes is "measured" on a scale from, say, 0 to 1. When this information has been collected, it is fairly easy to calculate which decision provides the highest average utility, the assumption being that the clinician must "maximize" the utility of his actions.

It is interesting that most decision theorists who advocate such procedures do not demand that the probabilities that enter the analyses are based on recorded frequencies. They accept the axiom of Bayesian statistics, namely, that a probability is a subjective measure of belief. The cardiologist's statement that he is 90% certain that one of his patients suffers from myocardial infarction might, for instance, be included in an analysis. However, published examples of decision analyses are usually based on frequencies found in the literature or frequencies estimated by experienced clinicians.

In principle, I fully accept the idea that a probability is a subjective measure of one's belief, but one must not ignore that decision analyses based on this probability concept are of a very personal character. The clinician who is faced with a difficult problem may have to assess the probability of different outcomes and in want of exact information in the literature he may have to rely on his uncontrolled experience from those cases he remembers, but that, of course, does not mean that such uncontrolled experience is reliable. Papers published by clinical decision theorists sometimes leave the impression that carefully controlled therapeutic and prognostic studies are really a waste of time as one may just as well ask a few "skilled" clinicians to state their vague, uncontrolled experience in numerical terms. This is, of course, not true. We once asked 143 doctors to assess the long-term prognosis of duodenal ulcer patients [6]: Some thought that 80% would become symptom-free after a number of years, whereas others thought that this would only be the case in a few percent of cases, and it is easy to imagine the effect of such variation on the reliability of decision analyses. A decision analysis may sometimes be useful in concrete clinical situations when the clinician has to make a decision on the basis of available biological and empirical knowledge, but statistical decision theory must never be used to gloss over the gaps in that knowledge.

The assessment of the utility of different outcomes of different decisions is even more complicated. Sometimes the unit of utility is survival years or economic costs. Decision analyses using these "hard" utility measures present few theoretical problems. There can be no doubt that in principle it is possible to calculate which decision provides the longest average survival or the cheapest management of the patients. However, serious problems arise when, for instance, wagering techniques are used for assessing the patients' preferences for different outcomes, as such procedures suggest that clinical decision theory allows not only for empirical and biological thinking but also for humanistic thinking. That suggestion, however, must be considered more carefully.

The literature on the assessment of utilities is considerable but I have yet to see a single paper that shows that it is possible for patients in real life to assess the utility of possible outcomes of different decisions in numerical terms. Clinical decision theorists are, of course, fully aware of the problem. They suggest as possible solutions that the family physician, who knows the patient well, may assess the utilities on the patient's behalf or that research is done to analyse decisions for classes of patients as opposed to decisions for individual patients ([15], p. 222). In both cases, however, the statistical decision analysis gives clinical decision-making a paternalistic twist.

From a philosophical point of view, clinical decision theory, as it is used in practice, is closely related to classical utilitarian moral philosophy. When we have the choice between different courses of action, we must select the one with the best consequences, and the goodness of the consequences must somehow be determined by empirical inquiry. In this way, decision theorists, like other utilitarians, try to replace that which I have called "humanistic thinking" by "empirical thinking".

Pellegrino discusses in detail the complexity of value considerations in everyday clinical practice. In this context, I wish only to add that biological and medical decision-making which is based exclusively on biological and empirical thinking logically leads to a paternalistic attitude toward the patient. If the doctor takes it for granted that concepts of illness and health can be defined in biological terms and if he believes that the goodness of his actions can be determined by empirical research, there is no need to assign the patient an active role in the decision process. Of course, no clinician denies that medicine is both an art and a science. The scientifically minded doctor may well be kind to his patients, tell them about their diagnosis, and explain in detail which treatment is "best", but to him scientific and humanistic medicine constitute two separate spheres. He regards scientific medicine as self-contained and his decisions *qua* practitioner of scientific medicine do not require an understanding of the ways in which his patients relate themselves to their illness. As Kierkegaard puts it, "the way of objective reflection makes the subject accidental, and thereby transforms existence into something indifferent, something vanishing" [8].

If, on the other hand, the clinician takes into account that his patient is not only a biological organism, but also a self-reflecting human being, he will come to the conclusion that the concepts of illness and health cannot be contained within a naturalistic framework of thinking. Illness may well be regarded as abnormal biological function, but function is always function for a *telos* or purpose; the normality of biological function in a human being must always be related to the telos of that person's life, which reflects the person's self-chosen values.

In former days, doctors were mainly concerned with treatment of infectious diseases and other problems that could be explained and solved almost exclusively by scientific thinking, and they may have felt with some justification that philosophical discussions about the nature of man had little to do with everyday clinical medicine. Today, however, many of those patients that are seen by clinicians in Europe and North America present problems of a very different kind. The patients suffer from the degenerative diseases of old age, malignancies that can be treated but not cured, the effects of alcoholism, and so-called psychosomatic disorders. In all such cases, humanistic thinking is as important as biological and empirical thinking. The methods to be used are not only scientific but also hermeneutic, and the art and science of medicine can no longer be separated. A few years ago it was found that 25% of the patients who were admitted to the medical unit of a Danish teaching hospital presented what was called "significant ethical problems" and the subsequent analysis revealed that most of these ethical problems could be classified as "quality of life" considerations in connection with the management of patients with the diseases of old age or with incurable malignant diseases [9].

So far I have only discussed three types of computer-aided medical decision-making: branching logic, Bayesian probabilistic methods (i.e., conversion of conditional probabilities by means of Bayes's Theorem) and statistical decision theory. In his paper, Schaffner discusses recent developments in computer diagnosis, especially two programs called MYCIN and CADUCEUS (the latter being developed from INTERNIST-1 and INTERNIST-2). I have, of course, no personal experience with these programs, but from what Schaffner writes, the components are in principle those which I have discussed already. INTERNIST-1, for instance, just like branching logic, imitates the clinical thinking of its originator, and to this extent the user of the program asks for the advice of a competent colleague (cf., Schaffner's comment: "In a very real sense the INTERNIST-1 program is an AI-simulation of Jack Myer's clinical reasoning and his own internal knowledge-base" ([4], p. 208). The programs do not convert conditional probabilities by means of Bayes's Theorem, but Schaffner points out that INTERNIST-1 and CADUCEUS "reason" probabilistically, taking into account $P(S/D)$ and $P(D/S)$ of different diseases and disease manifestations. Therefore, these programs do not avoid the difficulties discussed in connection with Bayesian probabilistic diagnosis. I was, of course much impressed by the fact that it proved possible by means of INTERNIST-1 to diagnose Hodgkin's disease in the case reported in the *New England Journal of Medicine*, but I also noticed that the test case was one, where the pathologist was able to establish the truth of the diagnosis by independent means. Proponents of computer diagnosis usually use this type of disease to illustrate the capabilities of their programs, and I still do not see how they tackle all those diseases that, as explained above, cannot be diagnosed with certainty. Is it considered a success if the computer in those cases arrives at the same (sometimes somewhat arbitrary) conclusion as a skilled specialist at a well-known teaching hospital?

Finally, the programs do not only reflect empirical and biological thinking; they also include value judgments. For example, it is mentioned explicitly that investigations requested by the computer are ranked according to expenses and risk. The originators here took into account the consequences for the patient of overlooking different disease when in CADUCEUS they decided upon the "impact" of different disease manifestations. However, those value judgments that have been incorporated into the programs are those of the originators and not those of the patients. For this reason, computer diagnosis of this kind may promote a paternalistic attitude to the patients.

Since this comment and the two quoted papers are concerned with the *future* of medicine, I shall summarize my arguments by listing some of the needs that must be fulfilled in order to ensure continued progress in medicine.

(1) There is a need for more biological knowledge concerning disease mechanisms and environmental causes of disease. This knowledge is prerequisite for the development of more effective preventive, therapeutic, and diagnostic methods. In this connection, it may be mentioned that the importance of computer diagnosis diminishes when new effective diagnostic methods are introduced. In his book, *Introduction to Medical Decision-Making* ([11], pp. 59-63), which is now a classic, Lusted discusses the Bayesian probabilistic approach to the diagnosis of gastric ulcer and records the frequency of a large number of radiological findings in benign and malignant ulcers. Today, however, this approach has been made obsolete by the introduction of flexible gastroscopes, as it is now possible to take multiple biopsies from the edge of the ulcer.

(2) We also need more empirical clinical knowledge about the frequency of symptoms and signs in different diseases and about the short-term and long-term prognosis of different categories of patients. The collection of such information requires meticulously planned prospective studies. It must also be stressed that the introduction of computers can never replace the need for empirical testing of new treatments and diagnostic methods by means of carefully controlled clinical trials. Whenever information from such trials is wanting, which is frequently the case, "ordinary" and computer-aided decision-making must be regarded as equally unreliable.

(3) It must be hoped that the present wave of interest in medical ethics and humanistic thinking will continue. I believe that clinical decision theorists would regard their efforts more critically if they knew more about moral philosophy and realized the similarity between modern decision theory and the unsuccessful "felicific calculus" which was suggested by the utilitarian philosopher, Jeremy Bentham, at the beginning of the last century. I do not suggest that utilitarian moral philosophy is obsolete, but those who defend utilitarian thinking must at least realize the philosophical implications of their position.

In addition, appreciation of the humanistic component of medical decision-making ought to make doctors realize that hermeneutic methods are as important in medicine as in, for instance, sociology.

(4) Clinicians must learn to take a much greater interest in clinical theory. At present, fundamental biostatistics is being taught at most medical schools, but the average clinician does not know enough about such topics as inter-observer variation and critical assessment of diagnostic and therapeutic methods. The terminology of clinical theory has not even been properly standardized. As illustrated by Casscells, doctors do not agree on the meaning of the simple term, "a false positive rate" [2]. Medical teaching in the areas of clinical theory must be improved in order to make the medical profession ready for the computer age.

We must also analyze more thoroughly the clinical decision process as it takes place without the use of computers. I agree with Campbell who stresses that diagnosis is a hypothetico-deductive process and that we ought to teach medical students accordingly. He writes: "Good technique in history-taking, physical examination, and the choice of investigations are dominated by the need to generate hypotheses quickly and to test them critically rather than wasting time and money collecting information" ([1], p. 136). If Campbell's Popperian attitude to clinical decision-making is correct, one must be somewhat wary of computer programs that aim at establishing the final diagnosis. Rather, programs that assist clinicians to "generate hypotheses quickly" might be favored. A recent study on dyspeptic patients shows that the Bayesian probabilistic approach may be used for this purpose [3].

In short, we should continue exploring the possibilities of computer-aided clinical decision-making, but we should also discuss in greater detail what computers can do, what they cannot do, and especially what we want them to do. I should like to have INTERNIST-1 or CADUCEUS at my disposal next time I see a problem case like the ones reported in the *New England Journal of Medicine*. For the reasons mentioned in this comment, however, I do not envisage that such programs would be helpful in the majority of cases.

Unit of Medical Philosophy and Clinical Theory
Panum Institute
University of Copenhagen
Copenhagen, Denmark

NOTE

[1] The term *Bayesian* probabilistic diagnosis refers to the fact that frequential conditional probabilities are converted by means of Bayes' theorem. *Bayesian* statistics is something quite different. It is a statistical school of thought, according to which a probability is not regarded as a long-run frequency, but a subjective measure of belief. Modern decision theory, which makes use of that probability concept, is sometimes called Bayesian decision theory.

BIBLIOGRAPHY

1. Campbell, E.J.M.: 1976, 'Basic Science. Science and Medical Education', *Lancet* 1, 134–137.
2. Casscells, W., Schoenberger, A., and Graboys, T.: 1978, 'Interpretation by Physicians of Clinical Laboratory Results', *New England Journal of Medicine* 299, 999–1001.
3. Davenport, P.M., Morgan, A.G., Darnborough, A., and de Dombal, F.T.: 1985, 'Can Preliminary Screening of Dyspeptic Patients Allow More Specific Use of Investigational Techniques?' *British Medical Journal* 290, 217–219.
4. de Dombal, F.T.: 1978, 'Medical Diagnosis From a Clinician's Point of View' *Methods of Information in Medicine* 17, 28–35.
5. de Dombal, F.T. *et al*.: 1974, 'Human and Computer-aided Diagnosis of Abdominal Pain: Further Report With Emphasis on Performance of Clinicians'. *British Medical Journal* 1, 376–80.
6. Greibe, J. *et al*.: 1977, 'Long-term Prognosis of Duodenal Ulcer: Follow-up Study and Survey of Doctors' Estimates', *British Medical Journal* 2, 1572–1574.
7. Jennett, B.: 1979, 'Defining Brain Damage After Head Injury', *Journal of the Royal College of Physicians of London* 13, 197–200.
8. Kierkegaard, S.: 1944 (Original Danish edition 1846), *Concluding Unscientific Postscript*. Princeton University Press, Princeton.
9. Kollemorten, I. *et al*.: 1981, 'Ethical Aspects of Clinical Decision-making', *British Journal of Medical Ethics* 7, 679.
10. Koran, L.M.: 1975, 'The Reliability of Clinical Methods, Data and Judgments', *New England Journal of Medicine* 293, 642–664 and 695–701.
11. Lusted, L.B.: 1968, *Introduction to Medical Decision Making,* Charles C Thomas, Springfield.
12. Matzen, P., *et al*.: 1984, 'Differential Diagnosis of Jaundice: A Pocket Diagnostic Chart', *Liver* 4, 360–371.
13. Pellegrino, E.D.: 1987, 'Value Desiderata in the Logical Structuring of Computer Diagnosis', in this volume, pp. 173–195.
14. Schaffner, K.F.: 1987, 'Problem in Computer Diagnosis', in this volume, pp. 197–241.
15. Weinstein, M.C., and Fineberg, H.V.: 1980, *Clinical Decision Analysis*, Saunders, Philadelphia.
16. Wulff, H.R.: 1981, *Rational Diagnosis and Treatment*, 2nd ed., Blackwell scientific, Oxford.
17. Wulff, H.R., Pedersen, S.A. and Rosenberg, R.: 1990, *Medical Philosophy*, 2nd ed., Blackwell Scientific, Oxford.

EDMOND A. MURPHY

CRITIQUE OF DIAGNOSTIC FORMALISM*

I. OVERSIGHTS AND OVERVIEW

Drs.Pellegrino [12] and Schaffner [17] give admirable accounts of the state of the art of computer Medicine – scholarly statements from the bench, not the bar. Apart from asides, they grind no axes; they write with tempered enthusiasm, neither (as Pellegrino says) in romantic scorn of this interloper into august clinical practice, nor in uncritical adulation of this new *organum*. Both respect the collective, if unarticulated, wisdom of clinicians. Faced with such writings, the critic too easily turns either to lofty prose about something else, or to nitpicking, "talent snapping at the heels of genius".

Like Pellegrino and Schaffner, we medical geneticists too often inherit problems shallowly encumbered by hard men that reap where they have not sown. I feel for all those that must make the best of a situation they did not invent. They cannot write a balanced review and at the same time rebuild the edifice in terms that are apt to seem idiosyncratic. As critic, I have no such scruples; and my comments are not so much a criticism of these two exponents, as of the field they are called on to expound.

However timely computer diagnosis may seem in view of the massing of detail in medicine, it is logically premature in the sense that the foundations for it in clinical medicine are defective, as I shall try to illustrate. Worse, the inadequacies have inspired little research on fundamentals. There results a field much given to tactics, painfully little to strategy. I shall sketch another perspective repeatedly intersecting with the topics these two scholars have discussed.

But first some holes in the fundamental theory. Pellegrino states a conventional view that diagnosis puts the patient in a category "defined by some finite constellation of findings". Does he really mean *findings*? Is correct diagnosis tautologous restatement of the findings? If so, all misdiagnosis is merely inept. Or are some of the "findings" not in fact found until necropsy, if then? Later, he writes of "pathognomonic observations like a biopsy..." But given the definition, how can one prove that a finding is pathognomonic in the sense that the disease is its unique

255

J. L. Peset and D. Gracia (eds.), The Ethics of Diagnosis, 255–267.
© 1992 *by Kluwer Academic Publishers. Printed in the Netherlands.*

cause? If the Argyll-Roberston pupil is pathognomonic of syphilis, then *either* (on finite evidence) we deny that this pupil could ever occur except in syphilis, *or* the patient by definition has syphilis even if shown never to have been infected with *treponema pallidum*. This confusion is not Pellegrino's fault but an encumbrance from tacticians skimping on coherent theory.

Schaffner points out that the joint probability of two data is the product of their marginals if, and only if, they are independent. Otherwise one needs an exhaustive set of conditional probabilities. Now while that statement is true and clear in probability algebra, it is commonly neither in diagnosis. Statements about independence have no content *unless the sample space is defined.* Neglecting this point leads to odd paradoxes such as that the genotypes of the same sibs are at once both independently, and non-independently, determined [9]. This principle is a commonplace to epidemiologists.

Example 1. Sickle-cell trait and glucose-6-phosphate dehydrogenase deficiency are clearly dependent if the sample space is *All Americans*; for both are frequent in Negroes. Perhaps they are almost independent if the sample space comprises *All West Africans.*

Example 2. Diagnosticians often use data on consanguinity. But details of consanguinity may throw little light on a common phenotype in a mouse from an inbred line. In the inbred patient (e.g., from an isolate) what do details of consanguinity add? One expects him to be inbred, just as a prosector expects his subjects to be dead.

Schaffner carefull avoids this trap: in discussing dependence he adds "given the disease". But he adds in italics *"This is a condition generally violated in medicine."* I am disturbed at his glossing over the notion of "the disease". I can imagine two possibilities.

If the diagnosis is a *point* in the diagnosis space, the variation in its manifestations is pure random noise. If, as many suppose, there is no true noise, and hence no randomness, then stochastic independence has no meaning at all. If there *is* noise, the medical scientist must have some model of how random noise can be contagious. (I confess I find this unimaginable, smooth assumptions of the probabilists notwithstanding.)

If the disease is *not* a point but a *region* in the diagnosis space comprising an arbitrary category of fairly similar disorders grouped for con-

venience, then the dependence may be mainly (perhaps wholly) the facti-
tious dependence that results when homogeneous subsets are pooled. The
wider the pooling, the more false dependence is. Schaffner's position
may be saved if "disease" is to be a pure operational convention.

What is the empirical evidence for Schaffner's italicized comment?
And in interpreting it, how much attention has been given to the hidden
heterogeneity of the putative diseases? Commonly, the sample space used
is the whole population. But by no contortion can a patient be seen as a
random sample from that population except under some null hypothesis,
which, by asymmetry, violates Bayes' theorem.

The goal of diagnosis is perhaps best seen as an *attempt to define a
precise sample space of which the patient is a realization.* On such a
sample space (which must also embrace prognosis and therapy), the
findings will prove redundant (and fully dependent) or have low dependency, perhaps degenerately so. It would be of interest to explore the
diagnostic process as a search for a sample space on which all the
findings are either fully dependent or fully independent.

II. ANOTHER PERSPECTIVE

The vagueness of the nature of disease and the objectives of diagnosis
are to me much more disturbing topics than the purely formal problems.
No doubt there are many views about the goal of diagnosis. Some
perhaps think that the evidence is the point of the point: the goal is to
find and assess what overt perturbations the data exhibit. High blood
pressure is blood pressure that is high; and heart failure is the failure of
the heart to maintain an adequate cardiac output. My perspective as a
physician entails certain different suppositions, notably that diagnosis
proper is a non-trivial search for something, *the sought*, distinct from *the
seeking*, i.e., the diagnostic process and the use of evidence. This distinction will be repeatedly invoked.

By *formalistic diagnosis*, I mean the belief that clinical diagnosis,
prognosis, and management can be turned into an assembly of explicit
and logical steps. This thesis proposes (as artificial intelligence does) to
substitute, without any loss or distortion, even with some refinement, a
calculus of formal mainpulations for *de facto* clinical medicine.
Proponents may point to precedents in other sciences, though recognizing that medicine is complex, impeded by intellectual inertia and practical demands. Their appeal contains much to be applauded; and I do not

deny its achievements. Nonetheless, in casual comments and occasional, even authoritative, writings [2], physicians have judged formalism interesting but inadequate and overexplicit; and we must wonder why. It is fruitless to ask those that belittle explication to explicate their reasons. However, the critics, to command respect, must do better than sneer at the centimeter-gram-second system.

A critique of formalistic diagnosis may address two main issues. First, are the logical steps sound and the empirical data reliable? I have little to add to the vast literature on this topic, so ably reviewed by Pellegrino and Schaffner. Second, is the formalization adequate, comprehensive, correctly weighted?

III. PROBLEMS OF THE DIAGNOSIS SPACE

A. Goals

Neither reviewer addresses the question of whether the diagnostic process (however much it may vary in detail) has uniform goals. I believe that it does not. If *hyperglycemia* means merely that the blood glucose is above 180mg/dl, then given the biochemical fact, there is no call for subtlety. That kind of diagnosis is pure tautology. If it is not, then the definition is inadequate. However, prescinding from issues of its form, we may take it that the goal of diagnosis is at least *sometimes* a conclusion that is more than a restatement of the evidence. Such diagnosis implies some interpretation, processing, or setting-in-context; and at least some diagnosis is seen (rightly or wrongly) as a search, using clues, for a location in a *space* (or set) of diagnoses. The point sought must be called something and we shall use the code term *correct diagnsis*.

B. Resolving Power

This is a property of the seeking. The data are mostly noisy, because of ambiguity of language, faulty memory, reticence, and what the diagnostician and the patient think relevant. There are practical, perhaps theoretical, limits to resolving these flaws, which are often disastrously confused with random error and treated as such [8]. Beyond a certain level, distinguishing an odd feeling in the chest from discomfort, or discomfort from pain, is, unlike some parameter of a random variable, incapable of being refined by asking the patient more and more questions.

Example 3. One may cite the male patient who complained of passing green urine [13]. Now certain vegetable dyes produce green urine so the complaint is not absurd. However, further questions established that he is colorblind, but throw no light on what the color of the urine had actually been. (He had hematuria).

C. Properties of the Search

Certain assumptions, not clearly warranted, are made about the properties of the diagnostic search. They should be at least systematically doubted; and three in particular are of some practical importance.

Uniqueness. This is a property of the sought. It is assumed that there is only one correct diagnosis. This statement must not be confused with the purely evidential statement that one can find only one answer consistent with the facts. Rather, it is akin to supposing that an equation has only one real root, which is quite a distinct issue from whether we can find it correctly. At times, this assumption must be in doubt.

Example 4. The psychiatrist may be unsure whether a patient's viewpoint is a distorted ideation or an extraordinarily subtle insight. (The two are, of course, neither exhaustive nor mutually exclusive.) This dilemma is dependent on culture.

Convergence is a property of the seeking. It is assumed that in the diagnostic space, sufficiently close to the correct diagnosis ("within the zone of convergence"), there exist assured pathways to more accurate diagnosis, which are limited only by the resolving power. Formally, the evidence (e.g., likelihood) must be non-concave, so that diagnostic truth can be pursued by systematic adjustments as in iteration. But the space may not be continuous.

Example 5. The cytogeneticist determining the ploidy of a tissue is constrained to give an answer that is an integer. The likelihood is then not continuous and therefore not differentiable. Moreover, some well-know functions though everywhere continuous, are nowhere differentiable [6]. This problem may be dismissed as too formalistic. Nobody complains that the (true) mode of a Poisson variate cannot be found because it does not meet the operational criteria imposed in finding the maximum by differential calculus. But it leads at once to a deeper problem.

Well Behavior is a property of the sought. Granted that there is a unique answer to be sought, and granted that evidentially it is the most plausible, the common assumption is that the next best conclusion is arbitrarily close to it. Thus Newtonian mechanics, an earlier approximation, results from special relativity when we suppose that the speed of light is infinite. Then the relativistic model is not something totally other than Newtonian mechanics, but a refinement of it. Newtonian mechanics works because for most practical purposes the speed of light is well approximated by infinity. This relationship between closeness to the truth, and the aptness of the approximation is *well-behaved* if there are no local pathologies in the space of interest. It implies that, if the diagnosis proves to be not quite right, we should search in the same neighborhood and not in some totally unrelated part of the diagnosis space. To the formalist, this is a strong assumption.

Example 6. What is the sign of the maximum absolute value of $[sin(1/x)]/x$? It is undefined at zero, but within any neighborhood of zero, there are a countably infinite number of extreme values, of alternating sign and mounting magnitude. More broadly, the closest rival to the truth may be more apt than a perturbation, however minute, from the truth.

Example 7. A patient with bronchiectasis came to the emergency room saying that he had coughed up blood. Since confirmatory signs of bronchiectasis were present, one is apt to think he should be admitted to hospital. The doctor may not be convinced that there has been hemoptysis if the history is vague; but he may not be convinced if the history sounds too perfect, as if the patient has read it in a textbook. Now suppose we took seriously (as I am sure Pellegrino did not) the proposal that the patient should make his presenting complaint unaided to the computer. If conviction is a well-behaved function of evidence, how is one to *formalize* the perception that the history is "too good"? Does the history have to meet criteria of both badness-of-fit and goodness-of-fit, as Fisher seems to have required of Mendel's classical experiments [4]? Can an otherwise convincing history be wiped out by one absurdity?

Example 8. A patient gives a history of attacks of "cold sweating". It is a classical sign of secretion of *nor*adrenaline; but most diagnosticians would not think that an adequate diagnosis. To focus on one line of thought, the patient may have a pheochromocytoma; in that case the blood sugar may be high but calls for no special treatment. Alternatively, the patient may be reacting, with secretion of *nor*adrenaline, to hypoglycemia due to starvation or an overdose of insulin, in which case the blood glucose will be low and the right treatment is to give carbohydrate. But also the hypoglycemia might be a reaction to hyperglycemia as occurs in the dumping syndrome; or to overshoot due to a prolonged lag in the homeostatic response. In these cases, giving carbohydrate merely aggravates the problem. It may be due to hypothlamic impact on the lag time that in turn might be compensatory [15] and the glucose *load* should be limited and carefully redistributed. It may be due to demands of the thalamus or higher levels in which case the treatment is obscure. The further one traces this complex chain, the more ambiguous the relationships among cold sweats, blood glucose levels, and management. The disorder may be quite unrelated to the *mean* level of glucose in the blood. There is a wild oscillation in the relationship between the blood glucose and the cold sweat that implies that the properties of the diagnostic space verge on the incoherent.

Now, of course, if the program is merely a shapeless *ad hoc* check-list of the causes of cold sweats, no doubt the computer could be programmed to find an answer; the only snag is that the particular answer is limited by the experience of the programmer. The computer then, like a telephone directory, is a vulnerable aid to memory, with nothing like the rationale of a sentient diagnostician. Rational approaches are more demanding.

Example 9. In the film *The Ruling Classes*, there is portrayed a man with a crass delusional insanity who on treatment becomes cool, clear and rational; and this state uncovers a deeper, more elaborate, more sinister, state of mind, much less clearly insane because it embodies certain cultural values that are themselves open to the charge of being distorted. The story is a cruder version of the subtle ambiguity that underlies *Hamlet*. Much is written on how, as we peel layer after layer off the evidence, we endlessly oscillate between the idea that Hamlet is mad and that he is not. The progression is not at all like a smooth and mounting assurance that he is, or is not mad. Of course, *Hamlet* is not a psychoanalytic account; but it lays bare ambiguities in our habitual notion of sanity, which is not entirely a psychiatric issue.

IV. PROBLEMS OF DIMENSIONALITY

Orthodox Bayesian analysis and discriminant analysis, if controversial, are well established, and I shall not deal with them. Bayesian diagnosis is the more flexible and more demanding. There are limitations to both, as the sterner problems of cluster analysis and numerical taxonomy bring to light.

A. The Economy-Fidelity Tension

It is commonplace that as parameters are added, whether *ad hoc* or not, the resulting fit to the data improves. Even if the model meets reasonable criteria of goodness of fit, there is still unease that high parametricity means evidential shapelessness. Poisson said "Give me four parameters and I will produce an elephant; give me five and I will make it wave its trunk". A snag with diagnosing such highly pleomorphic diseases as polyarteritis nodosa and sarcoidosis is that they explain too much too painlessly. From an extensive experience of multivariate, non-Gaussian data, it is known that the set of regressors deemed significant is often sensitive to the order of the crossproduct. It is also disturbing that in non-Bayesian analysis (notable step-wise regression) the "significant" factors are commonly colored by *the order in which the analyst thinks of them* [7].

B. Nosological Multiplicity

It is well-know [3] that unconstrained fitting of a mixture of continuous distributions to a set of data always gives in the limit an infinite likelihood. The bestfitting solution is one that puts each patient in a different class, a process close to taxonomic nihilism, even nominalism. I reserve judgment as to whether that apparently absurd conclusion may after all be correct. Now diagnosis is a somewhat different problem; but there are analogies.

Example 10. On routine examination, the patient had pupils that did not react to light; an absent right ankle jerk; and paraesthesia in the anterolateral aspect of his left thigh. In consultation, a distinguished and punctilious neurologist who could find no other signs whatsoever, reluctantly concluded that the patient had neurosyphilis, an old slipped disc, and meralgia paresthetica: three signs; three diagnoses. The point is not whether these diagnoses were correct; it is that experts (the class who, at present must be taken as the gold standard of diagnosis) should see fit, however reluctantly, to have the same dimensionality for the findings and the diagnosis. In the statistical analogue that leaves no "degrees of freedom for error". Little more noise would lead to algebraic indeterminacy and disrupt the formalistic approach.

As we increase the number of diseases that may be ascribed to one patient, we convert the original diagnostic space into a cross product space; and since we have no right to suppose the diseases probabilistically independent over the sample space of the population, and total independence cannot be inferred from marginal independences, there seems to be no practical limit to the details needed to equip all these cross-product spaces. If a patient has ten findings, what is to stop a computer from making ten diagnoses? On the one hand the prior probability of ten different diagnoses may be small; on the other hand, the likelihood of each may be indefinitely large so that their product is very large. In principle the diseases may even outnumber the signs; but diagnostic extravagance is often condemned as distasteful, and it makes management confusing.

Computer diagnosis might avoid these issues by being given a limit on the order of the cross-product space; but how would we decide the limit, or what logical contortions would we tolerate to keep the diagnosis within it?

V. PROBLEMS AT AND BEYOND THE BOUNDARIES

A disturbing feature of recasting clinical diagnosis in terms amenable to computer analysis, is that (tautologously) a system must impose structure, which may at times be either too lax or too rigid. No great diagnostic finesse is needed in deciding that a man whose hand has been accidentally put in boiling water has been scalded; and an excruciating family history (such as a rigid computer program might demand) is neither here nor there.

Shaffner recounts two main approaches, *Bayesian* and *branching*. I mistrust this specious dichotomy. Bayesian methods may be sequential, but unlike branching logic, not dynamic. I cannot see how they show *what datum to seek next*. In my view [8], neither method prescribes that the density of search vary (in the sense that in the Newtonian method of iteration the size of the step changes as we near the maximum). Nicety in exploring a symptom increases as the diagnostician narrows down the system involved.

However, it is when the diagnosis demands relaxation of a constraint that the greatest difficulties lie. To identify such shortcomings in a system is not to say that they are forever incurable; but if formalistic diagnosis is to replace classical, then there must be ways for the system to get outside itself and make adjustments. A rough analogy with Russell's principles of orders of language [16] suggests that this step is *logically* impossible.

A. Articulation

Arguably, formalistic diagnosis sets an undue premium on what can be articulated. All clinicians use the *Gestalt* in its correct and aboriginal sense: the perception that depends on integration of components that severally are imperceptible. The classical example is the gait in Parkinsonism that may be undetectable on even the closest scrutiny of the parts. But a much more complex issue arises when articulation is not merely difficult but perhaps, by nature, impossible. Incoherence and randomness, are two states often, and disturbingly, confused. One main practical criterion is that the behavior of a random process converges in probability to a distribution function, whereas that of chaos does not. In applying this criterion, we may be frustrated by secular changes in a random process that prevents us demonstrating such a convergence. Diagnosis of certain conditions depends on the fact that they seem to be incoherent.

Example 11. Choreic movements are purposeless and unintended. Like voluntary movements and unlike tics, they are not repetitive and patients learn to make them look voluntary. Usually the disorder improves; and even if it is random, the parameters change over time; so in mild cases, there may be too few movements to see whether the process converges on a distribution function. With these unsatisfactory properties, it is hard to say how the diagnosis is made or *a fortiori* how a program can aim to diagnose it.

B. Limits

Diagnostic formalism may require that the diagnosis space be of agreed form and based on adequate knowledge. It may work well most of the time; but it makes no provision for missing knowledge. Shortcomings may be made good by synthetic translations. If a patient has an unexplained high blood level of threonine, the state may be translated *idiopathic hyper-threonin-emia*. However, clinicians do meet *new* diseases, and resisting the lore of empty translation into bogus Greek, they seek, and may find, what is wrong and what are its implications.

Example 12. From one case of oroticaciduria, its mode of inheritance and management were inferred [5]. By itself, a precast computer program could not have taken these steps.

C. Ephemeral Modifications.

A clinician seeing three cases of influenza in one morning, surmises threre may be an epidemic, and temporarily changes his prior probabilities. A computer can do the same; but this logical contagion raises the estimated frequencies *even if the disease is not infectious* [10]. The fallacy lies in using a decision as a scientific datum [8], which in turn is due to confusing a private hypothesis (personal diagnosis) with a public hypothesis (an empirical relationship defined on a reference population).

Example 13. Campbell [1] suggested that the mysterious rise in the incidence of coronary disease in this century is an artifact produced in this fashion. He may have overstated his case somewhat. In recent years, coronary disease has declined [18]. Would he call this pattern a counter-contagion?

VI. PROBLEMS OF MEANING

Several lines of thought persuade me that obsession with formal proof has stifled the role of meaning in rational conviction. In didactic medicine, it is now grossly neglected, partly because of the shallow perception of medicine as applied science, ignoring its personal component. *Meaning* was long since schematized and adorned with much awesome vocabulary ([16], [11], [14]); I limit myself to simple lay uses of the word in diagnosis. It is a major scotoma in diagnostic formalism. A person is unique; so any probability statement about a patient lies within subjective probability, not empirical statistics. Laws of large numbers and the limiting-ratio idea of probability are inapplicable. A probability is no longer a proportion: To say "60% of this patient is diabetic" is meaningless. (Such statements *are* intelligible in a genetic mosaic when the sample space is not the patient but his body cells. It is then usable in assessing risks in the various parts of the body, e.g., of malignancy in neurofibromatosis.) Again, data may be converted into probabilities about patients by supposing that they belong to a natural sample space. But that surmise is perilous and at times (e.g., in fingerprints, which are unique) manifestly false.

Example 14. The estrangement between formalism and medicine is illustrated in psychoanalysis, which assumes that each patient has a unique somatic endowment, a unique set of experiences, a uniquely structured unconscious, and idiosyncratic symbols for expressing the content of the unconscious. In part, the psychoanalyst must crack the individual code, a harder task than cracking a literal code. Yet that solution only provides *coded meaning* such as that contained in DNA. If the unconscious were to communicate in plain English, there would still remain the issue of *implicative meaning*.

Example 15. The details of the aura of an epileptic fit may show various levels of organization. When the patient has detailed, highly organized visual hallucinations, the computer may be programmed to find out if they are unilateral (provided the programmer thinks to include the point), which would suggest a temporal discharge. But who knows what other, more subtle, patterns might need to be sorted out?

VII. CONCLUDING COMMENT

The computer is a useful, perhaps indispensible, means of compiling data. If it is to assume any more of the clinician's role, then vastly more study must be done on fundamental knowledge about both the clinical and the mathematical aspects. They cannot be casually disposed of in exiguous footnotes. We have reason to fear what success in a modest experience may fuel in the zealot. My examples are not meant as mere sniping. I cite them to show that my criticisms are more than theoretical, and have practical importance.

Johns Hopkins University
Baltimore, Maryland, U.S.A

NOTE

* The work for this study was supported by grant GM34152 of the National Institutes of Health. The author is indebted to Drs R.E.Pyeritz, P.R. Slavney, and C.A. Rohde for valuable critisms and suggestions.

BIBLIOGRAPHY

1. Campbell, M.: 1963, Death Rate from Diseases of the Heart 1876–1959, *British Medical Journal* **2**, 528.
2. Degowin E.L.: and DeGowin R.L.: 1981, *Bedside Diagnostic Examination*, 4th ed., Macmillan, New York.
3. Edwards, A.F.: 1972, *Likelihood. An Account of the Statistical Concept of Likelihood and its Application to Scientific Inference*, Cambridge University Press, Cambridge.
4. Fisher,: R.A. 1936. 'Has Mendel's work been Rediscovered?', *Annals of Science* **1**, 115–137.
5. Huguley C.M. (*et al.*: 1959), 'Refractory Megaloblastic Anemia Associated with Excretion of Orotic Acid', *Blood* **14**, 615–637.
6. Minassian, D., and Gaisser, J.W.: 1984, 'A Simple Weierstrass Function', *American Math Monthly*, **91**, 254–256.
7. Murphy, E.A.: 1982, *Biostatistics in Medicine*, Baltimore. Johns Hopkins University Press.
8. Murphy E.A.: 1976, *The Logic of Medicine*, Johns Hopkins University Press, Baltimore.
9. Murphy, E.A.: 1979, *Probability in Medicine*, Johns Hopkins University Press, Baltimore.
10. Murphy, E.A.: 1981, *Skepsis, Dogma and Belief*, Johns Hopkins University Press, Baltimore. Especially Appendix.
11. Ogden, C.K., and Richards, I.A.: 1979, *The Meaning of Meaning*, 10th ed., Routledge and Kegan Paul, London.
12. Pellegrino, E.D.: 1987, 'Value Desiderata In The Logical Structuring of Computer Diagnosis', in this volume, pp. 173–195.
13. Platt, R.: 1965, 'Letter to The Editor', *Lancet* **2**, 1125.
14. Polanyi, M., and Prosch, H.: 1975, *Meaning*, University of Chicago Press, Chicago.
15. Renie, W.A., and Murphy, E.A. 'The Dynamics of Quantifiable Homeostasis III. A Linear Model of Certain Metrical Diseases', *American Journal of Medical Genetics* **18**, 25–37, 1984.
16. Russell, B.: 1950, A *n Enquiry into Meaning and Truth*, Unwin, London.
17. Schaffner, K.F.: 1987, 'Problems in Computer Diognosis', in this volume, pp. 197–241.
18. Stallones, R.A.: 1980, 'The Rise and Fall of Ischemic Heart Disease', *Scientific American* **242**, 53–59.

EUGENE V. BOISAUBIN

HUMAN VALUES IN COMPUTER DIAGNOSIS

I would like to begin with two examples of how human values might be added into the calculus of computer diagnosis and clinical management. The first issue concerns the implications for outcome when a particular diagnosis is made. Not all diagnoses are made in the same way. Some have significantly adverse implications if they are present and others do not. Take, for example, a case of a middle-aged man with an episode of hematuria following mild abdominal trauma. There are a number of possible diagnoses, each with an associated outcome. The first option is a condition with no adverse outcome for the patient, for example, the presence of a solitary renal cyst. The primary reason for making such a diagnosis is that the cause of bleeding is identified and other conditions do not have to be considered. The patient is not directly benefited but is protected from future concern and the cost and risk of further examination. A second possibility is a condition with a bad or even fatal outcome for which nothing can be done, for example, an advanced pancreatic carcinoma which has now invaded the kidney. A third intermediate possibility is that a condition is found that is serious and treatable, or even curable. An example of this would be an early resectable renal cell carcinoma. From the viewpoint of the patient, diagnosis number three is the most important one to make since it provides an opportunity to save the individual from a potentially fatal disease and restore health. A computer assisted diagnostic workup therefore should be directed towards making or excluding diagnoses such as number three – serious diseases that can be treated. One might argue that diagnosis number two deserves also to be made as a matter of priority because of the need to prepare the patient for palliative therapy and eventual death. But this situation should not receive the same priority as an urgent or emergent correctable condition. For a common condition such as hematuria, it should be possible to gain some consensus from the medical profession as to which conditions should be prioritized in the diagnostic scheme to provide maximum patient benefit. Indeed, some decision-making programs currently exist, for example, a pharyngitis protocol that focuses upon treatable conditions with potentially serious outcomes, i.e. streptococcal pharyngitis[1]. Many

J. L. Peset and D. Gracia (eds.), The Ethics of Diagnosis, 269–272.
© 1992 by Kluwer Academic Publishers. Printed in the Netherlands.

diagnoses in medicine, however, such as lupus erythematosis have highly variable clinical presentations and are not amendable to easy categorization as conditions with correctable outcomes. But by adding into the computer additional information about the lupus, for example the involvement of the brain versus the skin, the diagnostic and prognostic implications for the patient may be made more precise. Other conditions may have little import for the physician but great meaning for the patient. For example, a mild trichomonas infection might be considered by the medical profession as a non-serious cause of dysuria but it may have serious social repercussions for its victims since it is categorized as a sexually transmitted disease. Or, the diagnosis of early Alzheimer's Disease creates few treatment options for the physician but has major, potentially devastating implications for the patient and their family. If the medical system is to be truly responsive to the needs of the patient, then a shift from values important primarily to the physician to those also of the patient is imperative. It may be possible to collect important values or concerns from collective groups of patients with certain diseases and weigh them for relative importance. If certain values such as associated pain and suffering, functional incapacity or even economic cost can be listed and quantified with satisfaction, it might be possible to add these into the traditional computer calculus.[4] If some values prove too intangible and idiosyncratic, then the computer might simply encourage the physician to address and measure these issues with the individual patient.

A second example of coordinating human values into the computer decision making process involves patient preferences for treatment. Although most clinical decision analyses using computers have focused upon the diagnostic process, I believe the potential for utilization in therapeutics and management may be more important. As medicine has become more complex, specialists and subspecialists have proliferated and a patient with a difficult diagnostic problem is more likely to be referred to a knowledgeable, experienced specialist than dally with a primary care physician; with or without a computer. Second, the modern specialist is increasingly bypassing the traditional intellectual process of diagnostics by utilizing sophisticated technologies to diagnose or exclude immediately, a broad variety of conditions. As specialization and technology change the traditional diagnostic process, the computer's role in the increasingly complex area of medical management may begin to grow. Personal patient preferences for both diagnostic and therapeutic options will continue to grow as more choices are made available and patients are educated as to their existence. For many medical and particularly surgical conditions however, the doctors' and patients' options for treatment are limited. There are no real options for the treatment of diplococcal pneumonia except a penicillin-like drug and similarly no effective options exist for a sub-dural hematoma except surgical evacuation. But there are a number of conditions, both serious and non-serious, for which a variety of treatment choices exist without one having a clear-cut superiority. In these situations, personal patient preferences, particularly concerning the nature of the treatment and possible side effects could and should carry most weight in the decision process. The treatment of laryngeal carcinoma provides one example. Although total surgical laryngectomy has the best possible five year survival outcome, it also results in loss of voice – an intolerable situation for some patients. The other treatment option, local radiotherapy, offers a slightly reduced cure rate but allows preservation of speech. Studies have shown that patients have clear individual preferences about how much they are willing to extend minimal life as opposed to preserving important body parts or functions, such as the larynx, or reproductive organs [3]. Weighing patients' treatment preferences for a computer program may be initially difficult, but the increased awareness of the spectrum of treatment options will allow the creation of additional computer "branch points" for greater flexibility in clinical management and prognosis. In addition, a patient preference approach to therapeutics mandates greater patient participation in treatment.

Less dramatic but more common and clinically relevant examples are provided by current controversies regarding the treatment of hypertension. Hypertension is a pervasive public health problem affecting 60 million Americans. When untreated, it may result in stroke, heart disease, and kidney failure. Virtually all hypertensives can be effectively treated with oral medication although perhaps one-third of them will fail to comply or take their medicine as directed [2]. Modern pharmacology has made available to the physician and patient many new and different therapeutic agents to control blood pressure. A major reason for patient non-compliance relates to the side effects of the drugs that impair the user's quality of life. If a drug creates undesirable side effects such as dizziness, difficulty in concentrating or impaired sexual potency it is doubtful that the patient will continue to take the drug regardless of its effect upon blood pressure. Some side effects, such as chemical ones are less obvious to the patient and involve alterations of blood electrolytes or lipids that may also have important future implications for the patients. All antihypertensive drugs have side effects and different patients may want to accept or reject these drugs based upon these effects [5]. For example a young, recently married accountant would probably refuse a drug that impairs intellectual or sexual function. These side effects, however, may be less troublesome for an elderly, already demented nursing home patient. A modern computer program could present a listing of antihypertensive drugs, each with listed side effects and associated probabilities for their occurrence. The physician, with patient input, would make the best possible clinical choice based on efficacy, minimal impact on life style and potential for compliance.

In summary, the computer now offers a new world of technology assisted diagnostic possibilities for the clinician and patient. Although this technology has primarily focused upon the efficient diagnosing of traditionally medically labeled diseases, the implications for a more patient-oriented system are substantial. First, weighing or biasing the computer to preferentially make diagnoses with important outcome implications for the patient is feasible and necessary. Second, addressing patients' preferences for treatment will not only expand the scope of treatment possibilities but will increase patient involvement in their own care.

Baylor College of Medicine
Houston, Texas
USA

BIBLIOGRAPHY

1. Komaroff, A.L., *et al.*: 1986, 'The Prediction of Streptococcal Pharyngitis in Adults', Journal Gen. Intern. Med. **1**, 1–7.
2. Moser, M.: 1986, 'Treating Hypertension – A Review of Clinical Trials', *American Journal Medicine* **81** (6c), 25–32.
3. Pauker, S.G. and McNeil, B.J.: 1981, 'Impact of Patient Preferences on the Selection of Therapy', *Journal Chronic Disease* **34**, 77–86.
4. Tannock, I.F.: 1987, 'Treating the Patient, Not Just the Cancer', *N England Journal of Medicine* **317**, 1534–1535.
5. Williams, G.H.: 1987, 'Quality of Life and Its Impact on Hypertensive Patients', *American Journal Medicine* **82**, 98–105.

SECTION V

THE ETHICS OF DIAGNOSIS IN THE POST-MODERN WORLD

GEORGE KHUSHF

POST-MODERN REFLECTIONS ON
THE ETHICS OF NAMING

The attempt to model diagnosis according to the epistemological criteria of the pure sciences leads to an objectivization of the patient that undermines the clinical goal of addressing human need ([7], pp. 183–4). In order to do justice to humanistic concerns, there has been an attempt in bioethics to redescribe or "model" medical reality in such a way as to avoid the transformation of the patient into an object ([22], [25]). However, the problem facing any such attempt is formidable: The paradox of the history of medicine is that the humanistic goal of addressing the patient's needs has been in large part furthered by way of a process in which the patient has been increasingly objectified and depersonalized ([23], p. 96). The challenge facing bioethics is thus to model diagnosis in a way that does justice to both the epistemological concerns intimately connected with objectivity as well as the ethical concerns associated with regarding the patient as a subject. Instead of rejecting "objectivity", we must find a way to rightly place it within a broader context.

Marx Wartofsky has observed that the "progress" of medicine toward the ideal of science has been measured by "the degree to which human judgment is replaced by objectified algorithmic procedures" ([42], p. 268). This depended on a conception of science that "comprehends science and human rationality generally on the model of machine technology" ([42], p. 268; [41], pp. 213–14; [23], p. 95). Wartofsky criticizes this approach, arguing that it confuses "human reasoning and imagination" with the "means it has invented to increase its range and power" ([42], p. 269). Wartofsky undoubtedly is correct when he observes that the means of thought have been used to model the nature of thought. But it is important to note as well that thought is not some "thing" that is somehow independent from its means and understandable apart from them. This point is well exhibited in Stanley Reiser's account of the role that technology has had in determining the nature of medicine ([32], [33]). Sandra Harding exhibits the further extension of the mutual interrelation of thought and its means, by highlighting the symbiosis between socio-political structures and technology [16]. This recognition does not

275

J. L. Peset and D. Gracia (eds.), The Ethics of Diagnosis, 275–300.
© 1992 by Kluwer Academic Publishers. Printed in the Netherlands.

involve an acquiescence before a materialistic account of historical reason, but simply an acknowledgment that there is an intimate interdependence between human being and its means of self-realization. This is so, whether thought moves behind or ahead of its material conditions.

The question is thus not whether human beings will come to understand their activity on the basis of their material conditions, but rather what specific means of self-expression will be used to develop self-understanding. The model used of normative reasoning has been the "scientific", mathematical one; a model that antedates the machine technology of the 19th century by over 2500 years, and is probably rooted in the material conditions of accounting and surveying (geometry=land measure), which, in turn, have been shown to be associated with the technology of writing.[1] Our question today will be: what model can we use to understand the activity of medical diagnosis in such a way as to overcome the above-mentioned conflict between the literary, scientific model of normative reasoning and the ethical interest in respecting human dignity?

In order to overcome this conflict, our model will have to be more comprehensive than the previous one, integrating the old model into a new framework which both appreciates and relativizes the objectivity that is modeled on literacy. The "instrument of thought" that can provide such a comprehensive model of clinical medicine is one that has been given considerable prominence in recent philosophy; namely, *language*. But here we must be concerned with more than just the structures of literacy. We must also consider orality; even further, we must consider the relations between orality and literacy. In doing this, the intent is to further develop James Locke's contention that language mediates the description and understanding of medicine and that "medical knowledge...takes a linguistic form" ([25], p. 42).

In developing a linguistic model of medical diagnosis, we will do more than simply account for medical reality in linguistic terms. Many fields involve diagnosis. The psychologist, social worker, parole board, judge, priest, father – these are just a few role-responsibilities in which diagnosis is central. In each case one is confronted with phenomena that must be placed within a particular linguistic domain. The "placing" or "naming" involves a process in which one moves from certain signs (or symptoms) to some sort of affirmation about that which a sign denotes. Diagnosis is thus a question of semeiotics, a word that appropriately pertains to both semantics and symptomatology.

In broadest terms, an ethic of diagnosis will sketch those issues relevant in ascertaining the praiseworthiness or blameworthiness of any diagnosis. If we abstract from the particularities of a field-specific diagnostic process, then diagnosis can be developed in terms of signification (the sign function), and an ethic of diagnosis can be developed as an ethic of signification; or recognizing that the activity of the sign can be developed as naming, as an "ethic of naming". Such a general ethic, however, would not provide a comprehensive ethic of diagnosis in medicine. General deliberations on the philosophy of language (esp. the sign) need to come into a mutually constructive dialogue with specific deliberations on diagnosis in medicine. Concretely, in the case of medical diagnosis, this means that a matrix of interrelation needs to be developed between the philosophy of language and that of medicine.

In this essay I shall attempt to set up the conditions that enable the dialogue between philosophy of language (the general) and philosophy of medicine (the specific). The conditions are foreshadowed in the linguistic relation between semantics and symptomatology. In order to show this linguistic relation as a real one the following conditions are needed:

1. *Diachronic*: The history of diagnosis, in both its technical and ethical dimensions, must be developed in a way that parallels the history of the sign. In crudest terms, the history of the sign can be divided into a pre-modern and modern period. In the pre-modern phase the concern was with the correspondence between the thought signified by the sign and the essence of the extra-mental thing.[2] The focus was on the relation between sign, thought and thing and not between sign and other signs. In the modern period reflection on the sign took a pragmatic turn. Nominalistic and Empiricist reflection called into question the relation between the signified thought and the signified thing. Focus shifted from the extramental relation to relations within thought. In Peirce, who together with Saussure can be taken as the father of the modern theory of the sign, signs were viewed in relation to other signs. One then asked about the desired end of deliberation and developed the theory of the sign as the logic which best enabled the realization of the desired end.[3] Finally, on the border between modern and postmodern thought, there is a concern with intergrating the essentialist and pragmatic deliberations. This concern is well embodied in the orality/literacy studies of such diverse authors as W. Ong, H.-G. Gadamer, and J. Derrida. [4]

If the history of diagnosis can be shown to parallel the history of deliberation on the sign, then the logic of the development of one can be called upon to explain and further develop the logic of the other. In the first section of this essay I shall develop the history of diagnosis in a way that shows the parallel with the history of the philosophy of the sign.

2. *Synchronic*: The nature of diagnosis needs to be redescribed in terms of a linguistic process. In the second section of this essay diagnosis will be described in terms of writing; i.e. the function by which an oral word (the language of illness) is translated into a written one (the disease description).

Once these conditions have been met, linguistic deliberations can be put in medical terms and medical deliberations can be put in linguistic terms. The conditions are then present for a mutually constructive dialogue between the philosophy of language and the philosophy of medicine. This dialogue gives a general expression of the dialogue between the clinic and laboratory that occurs within medicine. The ability to translate the language of the clinic into the language of the laboratory and visa versa enables the developments in one field to benefit the other ([10], pp. 67–68). A matrix of transformation between medicine and linguistics thus offers promise for mutual benefit. It also provides a context in which one can develop a diagnosis of diagnosis, thereby addressing the conflict between the epistemological and ethical dimensions of medical diagnosis.

I. GENERAL REFLECTIONS ON
THE MODERN CONTEXT OF DIAGNOSIS

What makes a diagnosis "good"? This will be the question that an ethic of diagnosis addresses. One way to approach this is to begin by distinguishing between two meanings traditionally given to "good" ([21], p. 14):

1. *Technical meaning* – A good diagnosis is one that is *true*. Here "goodness" is regarded as a *question of fact*. In a technically good diagnosis the name given corresponds to the phenomena being named. It is assumed that there are differing things, each with its appropriate essence. When one names correctly then one sets forth the sign that signifies the thought that rightly represents the essence.

2. *Ethical Meaning* – A good diagnosis is one that is morally praiseworthy. Here "goodness" is regarded as a *question of right* (in the sense of moral propriety). In an ethically good diagnosis the process of diagnosis is prosecuted in a way that does justice to the reasonable expectations of an individual, society, and the medical community. Generally, one asks: did the physician (in the case of medical diagnosis) do all that reasonably could be expected in order to obtain a technically correct diagnosis? And did the physician properly respect the autonomy of the patient and the guidelines of his profession and society?

According to this framework an ethic of diagnosis will consider the factors that make a diagnosis *ethically* good. It will thus bracket the questions of truth (the technical meaning of "good") and restrict itself to questions of right.

Since questions of fact are separated from questions of right, it will be possible to have an ethically good diagnosis that is technically bad, and vise versa ([21], p. 14). The norms by which the technical and ethical are evaluated differ. In one case (the technical) the norm is the phenomenon itself. *Correspondence* between the name and the thing is the measure of goodness. In the other case (the ethical), the norms are the reasonable expectations of the patient, society, and profession.

This distinction between fact and right enables us to go back in history and evaluate the goodness of a diagnosis from two perspectives. From an ethical perspective one can consider the expectations of the time and determine the goodness of a diagnosis on the basis of those expectations. And from a technical perspective one can ask how well a given diagnosis appreciated the facts of the matter. In the former case, one judges diagnosis by the standards of its own time. In the latter case, one judges diagnosis by present-day standards (our grasp of truth is used in place of the phenomena and correspondence is judged by the relation between the present view and the one being evaluated). This two-fold approach to history enables a constructive relation of the past to the present. The ethical concerns allow us to consider a historical phenomenon in its own terms and categories (this is central to any modern historical study) and the technical concerns provide a basis for relating the history to today.[5]

There is, however, one very significant problem with the above deliberations. Although many patients and even physicians attempt to draw a clear distinction between fact and value, the distinction does not do justice to some of the more profound reflections on the meanings of disease and modern medicine. It rather reflects a "pre-modern" view which does not sufficiently recognize the obstacles to knowing truly.[6]

The developments that led to the modern understanding can be divided into two stages. The first stage is well represented by Thomas Sydenham.[7] In the nosographic medicine of Sydenham there was a marked skepticism concerning the ability of a person to know the real essence of a disease.[8] But, according to Sydenham, this did not hinder the concerns of medicine, which sought simply to benefit humankind.[9] In this context, the measure of the technical merit of a diagnosis is no longer judged in terms of the correspondence between the thought named and the essence of the thing. Medicine does not seek knowledge for the sake of knowledge, but asks with a view to action. Thus, the merit of diagnosis is judged by the degree to which it enables the treatment or prediction of the malady in question. A *technically good* diagnosis is now taken as a diagnosis that leads to a successful clinical encounter.

This pragmatic turn changes the whole framework of an analysis of the nature and ethics of diagnosis. When we previously asked "what makes a diagnosis good?", the process we were evaluating was fairly clear – or at least it seemed to be. One knew what diagnosis was, isolated it as the object of analysis, and then considered the factors that made it good. But now diagnosis is no longer an isolated event that can be judged independently of broader processes. In order to evaluate it, one must ask about the place of diagnosis in the broader context of medical care. The ethics of diagnosis must then be developed in the context of the ethics of the clinical encounter in general.

Sydenham assumed that the goal of medicine was well defined and free of the moral values of a particular individual. One practiced medicine within an objective framework. Diagnosis was best performed by the unbiased, objective observer ([35], pp. 684–687). When an individual introduced his or her values into the process of diagnosis (e.g. by way of hypothesis) then a technically good diagnosis is hindered ([35], pp. 681–684; see also [40], pp. 147–150). But as we move to the present-day understanding of medicine, it becomes increasingly clear that particular cultural values – both nonmoral and moral – are constitutive not only of the goal of medicine but also of the very nature of disease. There is an unavoidable "observer bias", which plays an important role in the construction of medical reality.[10] This raises yet additional concerns for an ethic of diagnosis. Now one must not only ask about the nature of a thing (the question of knowledge) or how one can best treat a given malady. One must go further and ask how medical reality *should* be constructed. The answer to this latter question will then provide a more comprehensive framework for understanding the nature, meaning and ethics of diagnosis.

In the second stage of modern medicine, one finds an integration of the essentialist and pragmatic approaches to diagnosis. This integration can be seen in the "critical dialectic" between clinical findings and the laboratory sciences.[11]

In the essentialist understanding, a technically good diagnosis was one which named the thing in accord with its essence. The problem with this approach is the same we find in the case of any attempt to verify correspondence between a thought and a thing: in order to provide a correspondence verification one must somehow stand outside of one's own thought and comprehend the relation between thought and thing. But the human knower does not have this privileged standpoint. Thus, the pragmatist abandoned the concern with the relation between thought and thing; between signifier and signified. Sydenham and those following him came to view a technically good diagnosis as one which records the data of sense without the intrusion of hypothesis or any other observer bias. The merit of the diagnosis is determined by how well it enables prediction or treatment. The real essence – the signified of the symptoms, taken as signs – is either unknowable, transnational, or irrational ([6], p. 54; [35], p. 689). Thus, concern with the signified of the sign is abandoned and one seeks to live within a differential system of signs. The advantage of the pragmatic approach was that it kept the goal of medicine in mind and sought to formulate its theory in such a way that the concrete concern with healing was advanced. But it also had a significant weakness. To put it in the words of Sanchez-Gonzales:

It was never possible to make a successful translation of the propositions concerning physical objects to propositions about sensory data. This failing, in combination with other factors, led to the final admission that every sensory observation has theoretical and historical concent...All this has consequently begun to bring about the general questioning of the myth that scientific knowledge has an objective grounding ([35], pp. 648–685).

Thus, some sort of structure is given to the data of sense observation. This means that there is necessarily an at least implicit affirmation about the nature of disease that is present in any observation. Observer bias cannot be avoided.

For the essentialist the thing itself provided a check on observer bias. But the question is now: given the skepticism about the real essence of a disease, how can one obtain a check on the observer bias of clinical observation that parallels the signifier-signified relation of essentialist diagnosis? The answer to this was given in the laboratory.[12] In laboratory studies, one had a new linguistic domain that paralleled that of clinical observation. An appeal by the clinician to the laboratory served the function of a correspondence verification of a diagnosis. Although the laboratory was given priority over the clinic in matters of knowledge (just as the signified is given priority over the signifier), the clinic was given priority over the laboratory in matters of practice. The ability to provide a clinical application to a laboratory theory provides a verification of the pragmatic merit (or usefulness) of the theory. Now a technically good diagnosis involves two medical dimensions: On one hand, one considers the diagnosis independent of the broader processes. The laboratory provides correspondence verification. This model is in many ways similar to the essentialist one. But on the other hand, one considers the diagnosis as constitutive of the more comprehensive process of prognosis and treatment. And one must evaluate how the constructed network of medical reality (with its moral and nonmoral values) conditions the understanding of diagnosis.

Thus far, we have been concentrating largely on that which makes a diagnosis technically good. In now moving to the ethics of diagnosis, it is important to remember that the very distinction between the technical and ethical merit of a diagnosis was derived in the context of an essentialist viewpoint. While the distinction will still have merit in the modern context, it must also be recognized that there is an intimate interrelation between technical and ethical concerns, between fact and value. One can no longer neatly separate the two. Moral and ethical considerations establish the conditions in which technical evaluations are developed, just as technical considerations establish the parameters of ethical evaluations.

The modern context of a technically good diagnosis raises new problems in the ethics of diagnosis. H.T. Engelhardt highlights the ethical problems raised by the restructuring of diagnosis according to the dialectic between the laboratory and the clinic:

This restructuring carried with it an ideology that discounted the significance of patient complaints. Patient problems came to be understood as bona fide problems only if they had a pathoanatomical or pathophysiological truth value. Absent a lesion or a physiological disturbance to account readily for the complaint, the complaint was likely to be regarded as male fide" ([7], p 183)

Engelhardt sees here an "error" which involves forgetting "the goals and purposes of medicine" ([7], p. 184). Implicit in this critique is the assumption that the first concern is the pragmatic one, and the essentialist concern (seen in the correspondence verification by the laboratory) should be secondary, and serve the clinical needs. Medicine is an "applied" not a "pure" science; "knowing truly" should be subordinated to "acting effectively". Thus Engelhardt emphasizes that "clinical medicine begins from and returns to the problems of the patient" ([7], p. 184). The "error" can thus be avoided by hierarchically ordering the pragmatic above the essentialist criteria of "technically good".

There is, however, another way to view the central role given to the laboratory. Instead of a "forgetting" of the goal of medicine, one can view the ideology as an attempt to set up the epistemological conditions of an effective *medical* intervention.[13] It is recognized that "clinical medicine begins from and returns to the problems of the patient", but it is further assumed that there are many problems that are not medical ones. The pathophysiological or pathoanatomical component is then taken as the condition of a *medical* problem. To use an example discussed by Ronald Carson, a teenage girl's refusal to attend school is indeed a problem, but it becomes a medical problem when the cause is a bad complexion.[14] The identification of the somatic component then makes it possible to medically intervene. Without such a component some patient's complaints will be taken as male fide *from a medical perspective* only because the complaints cannot be addressed from within the medical domain. If such a component cannot be found, then the physician may refer the patient to another healer, e.g. a psychologist or minister. In sum: one can view the restructuring of diagnosis according to the dialectic between the laboratory and the clinic as the establishment of the epistemological conditions of a technically good medical diagnosis, where the pragmatic concern is already factored into the meaning of "technically good".

Several problems attend this alternative to Engelhardt's interpretation of the dialectic between clinic and laboratory. First, it forgets that the lab findings are themselves a constructed, theory-infected component. The appeal to the laboratory thus does not provide an actual correspondence verification of a diagnosis, but only models the essentialist criteria by establishing conditions under which a certain type of observer bias can be checked. Edmond Murphy well notes that "it is deplorable to suppose that what we ordinarily regard as facts are elements of reality uncontaminated by means of observation, selection, and implicit interpretation" ([28], p. 287). The physician, by appealing to the laboratory, attempts to resolve what may be regarded as a "scientific controversy" about a given malady by an "appeal to fact". But if he thinks that the fact to which he appeals is somehow the "thing in itself" (Kant's "noumena") then, in the words of Murphy, he "displays a lamentable confusion about the whole epistemology of science – especially medical and biological science" ([28], 287). It is thus inappropriate to assume that the laboratory can serve as a final arbiter of when medical intervention will be effective, or to assume that observer bias is fully checked when laboratory findings corroborate a given diagnosis.

Second, due to limitations in knowledge, there will be maladies for which a pathophysiological or pathoanatomical truth value could be given in principle, but cannot be given in fact by the attending physician. Should we regard a complaint as male fide from a medical perspective, simply because knowledge is not sufficient to establish the conditions under which it could be verified? Here it is important for the physician to recognize two important things: (1) There are limitations to the physician's gaze, and of the technology that extends that gaze, and (2) The patient, by way of his or her experience, may have access to information that the physician may not be able to obtain otherwise. Stanley Reiser has argued these two points ([32], [33]), showing why greater weight should be given to a patient's account of the nature and significance of the illness, and he makes his argument not just on ethical grounds (respect the patient) but also on technical grounds (one can know things one could not otherwise).

Finally, and perhaps most significantly, the attempt to define the scope of medicine in terms of a dialectic between the clinic and laboratory does not provide us with a characterization that maps on to what medicine has done and continues to do. The physician has also served as priest, counselor, comforter, and friend ([6], p. 56; [22], p. 10). And in our present, pluralistic context, what medicine does and the line between medicine and other social institutions such as education, politics and religion will vary, depending on the particular worldview and context of the one who makes the definition and draws the line. Medicine will do different things for different people, and the expectations of both physicians and patients will vary ([7], pp. 185–195).

This last point raises issues that confront not just medicine but all aspects of society. In what has been called a "post-modern context", we live in a time of radical pluralism, where confidence in reason and science as arbiters of conflicting visions and values has significantly eroded [9]. It will thus be virtually impossible to come up with universally acceptable accounts of the world and of morality. Concretely, in the case of medicine, this means that there will be many ethics of diagnosis, not just one. The ethics of diagnosis will depend on the expectations that physicians and patients bring to the clinical encounter; on the moral frameworks used to evaluate the encounter; and on the descriptions of the "facts of the matter" that condition expectations and moral understandings of both the "ought" and the "can". And each of these will vary among individuals.

The problems involved in any attempt to provide a universally accepted ethic of diagnosis can be illustrated by the following question: To what degree does and should a physician's individual moral considerations come into the diagnostic process and influence the outcome of the diagnosis? Several essays in this volume on diagnosis have shown that individual values are involved in diagnosis. There is an unavoidable "observer bias". [15] But is this fact a positive or negative phenomenon? Should the goal be to minimize the role of personal values or should it be to recognize and embrace the role that personal values have in the diagnostic process.[16] One can also ask: to what degree is "observer bias" an expression of the general values of a society and medical community and to what degree is it an expression of the particular values of the physician? Perhaps a physician can to a great extent set aside individual, personal values and step into the observer bias of the medical community. Then one must ask what the observer bias of the medical community *should* be. A physician can then be regarded as one who puts on a particular type of observer bias that is established by the medical community. All these issues do not simply involve a description or prescription. The way observer bias is described involves an observer bias that constitutes the phenomena in a particular way.

In any attempt to answer these questions, individual values come into play. There is no "meta-level" from which to provide an ethic that does not involve the values of the one developing the ethic. The fact of moral pluralism thus provides a seemingly insurmountable obstacle to the attempt at providing a universally accepted ethic of diagnosis. The discussion of the nature of diagnosis must be prosecuted from a particular moral viewpoint. The essays in this volume are a good example of this point. The Marxist will take a very different approach to the ethics of diagnosis than a Catholic, Existentialist or Libertarian. The ranking of concerns such as economics, religion, theories of state and the nature of medicine provide different frameworks for developing the interrelation of professional, social, and individual concerns. Each perspective views the role-responsibility of the physician in a different way. That, in turn, governs the expectation that a well-informed patient should have in entering into the contract with the physician. And all these factors are central in the very description of diagnosis.

Recognizing that there will be an unavoidable observer bias in the diagnosis of diagnosis, I would like to suggest that the criteria for evaluating the merit of an ethic of diagnosis will be found in the ability of such an ethic to enter into a mutually constructive dialogue with the concrete concerns of a field-specific diagnosis. The essentialist and pragmatic dimensions of a technically good diagnosis can be generalized into criteria by which the diagnosis of diagnosis is judged. The essentialist criteria will be judged by the ability of an ethic of diagnosis to provide a description of an ethically good diagnosis that corresponds to the concrete needs and concerns of a technically good diagnosis. In this way, the factors of a technically good diagnosis will perform the "signified" function that the laboratory performs in the dialetic between clinic and laboratory. The pragmatic criteria of the ethic are then measured by the ability of that ethic to provide an account of diagnosis that overcomes the conflict between "scientific" and "humanistic" concerns that plagued previous accounts.

Up until now, I have in large part sought to be descriptive in my analysis of the ethics of diagnosis (recognizing, of course, that individual values are involved in any description). But now I shall attempt to provide a brief sketch of a prescriptive model of diagnosis. Because of limitations in space, I can only be suggestive. I will not be able to argue for the model. But I point to the above-mentioned criteria as the means by which the following sketch could be judged.

Briefly stated, I will attempt to describe the process of diagnosis in linguistic terms, providing a model that enables one to account for the objectivization of the patient in such a way that it accords with the broader ethical goal of respecting a patient's self-description. We have seen that the narrowing of medicine given in the dialectic of clinic and laboratory cannot be taken as a normative account of medicine as it has been and is practiced. But I will argue that it is valuable as a regulative ideal. The epistemological conditions established in the modern approach to a technically good diagnosis provide the conditions under which medicine ideally operates; namely, the conditions of intersubjectivity and of scientitic verification by an ideal scientific community.[17] A physician can be regarded as one who puts on the observer bias of the medical community, and this will be a bias that privileges the epistemological criteria of the ideal medical community. To some extent this will be expressed in practice as a narrowing of the scope of medicine. But this will be only a tendency, not a fixed norm. This bias, with its tendency to narrow the scope of medicine, will be seen among all physicians, because it is rooted in the "objective" orientation of medical practice. But some physicians will give it greater weight than others, depending on the particular values of the physician, the patient, and the moral communities in which they live. Although the regulative ideal will not define the nature and scope of medicine as it is practiced, it will provide a focal point around which physicians from different moral communities can organize their efforts.

Here the goal of medicine is to attain "objectivity" in a reconstructed sense; not the "objectivity" of 19th century science, which was concerned with a value-free account of reality, but a 20th century "objectivity", which is concerned with the epistemological conditions of effective intervention and intersubjective verification. Such a goal serves the clinical orientation of medicine. But it also has implications that run counter to some of the taken-for-granted assumptions of bioethics.

A central focus of modern bioethics is the physician-patient relation. It is recognized that the patient is a person, whose autonomy and self-determination should be respected. Thus the self-understanding and values of the patient are to be given a central role in the clinical encounter. It is for this reason that Engelhardt criticizes a notion of good diagnosis, which involves a disvaluing of the patient's experience of illness in favor of some laboratory verification. Diagnosis is not just a question of knowledge or practice (the two factors constitutive of a technically good diagnosis). It is also a question of respect for the patient, and thus the *relation* between the patient and the physician in the process of diagnosis.

When we combine the concerns of knowledge and practice with the consideration of the relation between physician and patient, then we come into a context where a new model is needed in order to enable a constructive discussion of the issues. I would like to suggest here that the "meeting of the worlds" ([23], p. 96) may provide such a model. In the physician-patient encounter, two linguistic domains – each with its own signs and structures – come together and interact. Just as the clinical and laboratory worlds meet constructively in the dialetic of diagnosis, so on a broader scale, the world of physician meets constructively with that of the patient. In the patient's world, a malady is experienced as "illness"; in the physician's world, it is experienced as "disease" ([23], p. 95). The diagnosis is then a process of transformation, in which the patient's malady is translated into the linguistic domain of the physician. Problems of knowledge and practice will be put in terms of intersubjectivity. And it is in the context of this encounter of worlds that the parameters of the ethics of diagnosis can best be developed today.

II. POST-MODERN REFLECTIONS

A recognition of medicine's tendency to become narrowly focused on the somatic component of a patient's illness experience leads many to call for the 'good physician" to be a more *holistic healer*; e.g. to address not just the somatic correlate of a loss of sight but also the alienating experience of blindness, should it come ([39], p. 119). This attempt to *expand* the scope of medicine runs counter to the attempt to *narrow* medicine's scope that we saw in the case of those who seek to actualize the scientific ideal. One can view this as a tension between the ethical and technical ideals of medicine. The opposing tendencies work together to bring about the advancement of medicine. The drive to expand medicine's scope motivates biomedical scientists to discover and create the conditions under which new maladies can be effectively addressed. The drive to narrow medicine's scope keeps clinical practitioners focused upon those conditions that can be addressed most effectively, and simultaneously motivates the physician to coordinate his or her activities with other healers in society, like the minister and family.

It is important to keep both of the opposing tendencies in mind. If one loses sight of the ethical ideal, then one comes to the dehumanization of the patient that so often is lamented in bioethics. If, on the other hand, one loses sight of the technical ideal, then the call to holistic healing becomes a call to minister to needs that go beyond the training and ability of most physicians. The doctor becomes psychologist, minister and friend. A few stop-gap measures are given to provide the air of authority in all matters and then the physician is to become the central healer of society. This puts new expectations in the mind of the patient and can only result in disappointment when the expectations are not met. Already, many complaints against the medical community involve factors that go beyond the physician's role-responsibility. By associating the physician with the objective attitude and appropriately placing his or her role-responsibility in society, more realistic expectations can be developed on the part of the patient, and the physician can be freed to more effectively concentrate on the areas most congenial to his or her expertise.

Bioethics has not sufficiently appreciated the ethical significance of the tendency to narrow the scope of medicine and promote objectivity. We have already seen that the objective attitude is a value that is advanced by the medical community. It is central to the criteria of a technically good diagnosis. But it is not just the value of a physician or the medical community. It is also a value of the patient. By means of the physician, the patient objectivizes himself. He wants to place his body – already objectivized in the experience of pain ([23], p. 98) – within the domain of "science", and thereby hopes that somatic wholeness be restored. The patient, for the most part, does not come to the physician for holistic healing.[18] He comes for a healing of the body.[19] Metaphorically, one could say that a patient gives his body to science – its objectivizing, catagorizing, totalizing stare – so that science may give back the body in renewed integrity. One could compare it to giving a car in the shop for repairs. One does not want, first and foremost, a holistic repair shop – one that deals with the emotional stress of having one's car break down. Instead, one wants a good mechanic.

This analogy may put the objectivization of the body in too extreme a way. Surely, we do not just have a body. We are body. But, following Drew Leder, I would suggest that in the experience of illness, the body is especially objectivized ([23], p. 98; see also [26], p. 85). Initially, by way of pain, this objectivization is an imposition. It is a symptom of illness. But in going to a physician, the patient attempts to positively appropriate the imposition. Just as an increased white blood cell count signifies a mechanism of the body to combat disease, so the objectivization of the self as body may signify a mechanism of the whole person to combat illness.[20] (The symptom becomes simultaneously an expression of the malady and of the attempt to overcome it.) One distances oneself from the pain that is now *in* the body.[21] And by way of the physician, one places the body within the linguistic domain of the scientific, medical world; a world where all occurs in terms of temporal, structurally interrelated causes. By the means of science, the physician restores integrity to the body. At that point, the physician-patient relation, viewed in terms of the role-responsibility of the physician, is dissolved. The psychosomatic wholeness usually follows in course.[22] But if it is problematic then it is cheifly the concern of the psychologist, minister, family or friend; not the physician.

Admittedly, the above discussion is an oversimplification. Healing of illness (the psycho-somatic healing) and healing of disease (the somatic healing) are not disjunctive and discrete. And many patients have an expectation that goes well beyond the role responsibility I have outlined. But the question here is whether there are contexts in which the scientific, objectivizing approach is appropriate as a focus of medicine, and this question is significant regardless of whether that approach can be fully actualized or whether it does justice to present expectations. At this point, I merely wish to suggest that when we are concerned with delineating a concept of medicine that will apply to all physicians in a post-modern context, it may be appropriate to restrict the role-responsibility of the physician to somatic concerns. Then the patient's expectations can be restricted to the expertise of the physician. If a person wants more than somatic care, he can go to another who is appropriately trained. This is not to say that it will be inappropriate for a physician to address needs in a more comprehensive way. One may even argue that in order to be a *good person*, the physician should address more than the somatic dimension of illness. But here we will distinguish between the ethical responsibility of all people (the good person) and those responsibilities that specifically apply to the physician (the good physician). All, for example, should seek to help a person who has recently become blind adapt to the new, handicapped mode to life. To treat the alienating correlate of a loss of sight is thus not uniquely tied to the physician's role-responsibility. But to address the somatic correlate of blindness – that is uniquely the physician's task.[23]

The nature of objectivization that takes place as the patient submits himself to the physician can be well outlined by way of an analysis of what takes place in diagnosis. As Drew Leder well notes, "the clinical encounter involves a meeting of two worlds, one defined in terms of the physiological diseases, the other through illness experiences" ([23], p. 96). Each world has its own linguistic domain. Diagnosis is a naming, in which the physician translates a name from the patient's domain – an ambiguous, imprecise illness-language, from the perspective of the physician – into the domain of the medical community. This naming involves placing and claiming a given phenomenon as a possession. The phenomenon is no longer an unknown intruder from another world. It is a known, named phenomenon; not a demon but a "disease".

This act of naming is, to use a concept from the philosophy of Levinas [24], "totalizing". It places the experience of another as a possession within my linguistic domain. The other is no longer the radical other. The "thou" becomes an "it"; no longer the subject of address, but only the object of concern. A central meaning of the respect for personhood involves allowing the other to name himself or herself. In this self-naming, the other is given the freedom of self-determination. The right to name oneself is thus a concomitant of autonomy. This raises the question: Under what conditions is it appropriate to intrude upon the self-naming activity of the other? One answer is: when the other gives himself or herself to be named. And this is exactly what takes place in the standard case of the physician-patient relation. The self the patient gives is his or her own body (that self the patient both has and is). In the act of submission, the body obtains a new name. Now it has two names, one within the language domain of the patient, the other within the domain of the physician. The difference between these two domains is well illustrated by the difference between symptomatic and disease descriptions of a malady, or in the difference between illness and disease.

Before further developing the nature of this activity of naming, it will be helpful to highlight an important distinction which can be illustrated by way of the following analogy: Consider what takes place if I hire a contractor to build a house for me. Initial discussions on price, blueprints, quality of materials, etc. take place in a dialogue between two more or less autonomous individuals. An agreement is reached. Then the contractor goes about his work. I may stand by and judge the contractor's work, making comments throughout the building of the house. But it is not *we* who do the work. He does the work *for me*. I am outside of the contractor's activity. His labor is a self-expression for which I agree to pay. And the relation is one of exchange. But with a physician, there is a two-fold relation: As with the contractor (or one who fixes my car) I am *outside* the physician-patient relation (viewed in terms of the role-responsibility of the physician). I encounter the physician as an autonomous other. But I am also *inside* the relation, as the *object* of the physician's activity. At one level, the physician works *for me*. But at another level, the physician works *on me*.

Bioethics has concentrated on the level in which the patient encounters the physician as an autonomous other. But this involves only those factors that are pertinent to any contractual deliberation. It does not do justice to the factors that uniquely define the role-responsibility of the physician. In the physician-patient relation, the patient does not just contract a work (= self-expression) that is independent of the patient. Rather, the patient contracts in such a way that he given himself to the physician as the field of the physician's activity. In the physician-patient relation, the patient will thus have two modes of existence. And the "meeting of worlds" will likewise be developed in two ways.

One model that seems to be especially appropriate for discussing the physician-patient relation in a way that does justice to the patient as a field of activity is the literary model. Edward Gogel and James Terry sought to highlight some of the descriptive and prescriptive possibilities of this approach in "Medicine as Interpretation: The uses of Literary Metaphors and Methods"[15]. The patient is regarded as text, and the physician as interpreter. This model is especially appropriate for several reasons. First, the distance associated with the objective attitude is also present in the interpretive act. It has been made a constitutive moment of modern interpretation theory.[24] But there is also a sound awareness of the role that individual values and "pre-understanding" have in any interpretation ([3], p. 55). One can thus develop the nature of medical objectivity in a way that does justice to both the valuing of the objectivity and the recognition that a "pure objectivity" (in the sense of a 19th century view of science) is impossible. Second, the literary model provides ample opportunity to further develop the play of signifier and signified that is involved in any detailed discussion of the movement from symptoms to disease. The linguistic relation between semantics and symptomatology can be developed as a strong analogy for better understanding the nature of the thought processes that are involved in diagnosis. Third, the relation between interpreter-text well highlights the reduced status of the patient at the object-level of the physician-patient relation. The interpreter is a full person. But the text, while having a powerful influence, is at the disposal of the interpreter. It is an expression of personhood. But it is not a person. Finally, the literary metaphor used to discuss the object-level of the physician-patient relation balances well with the oral metaphor that is often used to discuss the level at which both physician and patient are autonomous. The model of dialogue (oral model) works well in describing the "meeting of worlds" that takes place between two autonomous individuals. The literary model works well in describing the "meeting of worlds" that takes place between physician and patient when we focus on the physician's activity and role-responsibility. For this reason, I shall hereafter refer to the autonomous meeting as the *oral level* of the physician-patient relation, and the object-meeting as the *literary level*.

The oral and literary levels run parallel throughout the physician-patient relation.[25] The concerns of the oral level are not dissimilar to those one finds in many other fields besides medicine. They regard matters of interpersonal relations such as negotiations over cost and consent. But the concerns that more directly relate to the expertise of the medical profession – e.g. the interpretation of an abdominal x-ray ([15], p. 213) – are developed in literary terms. Here the patient has the status of "object", and the physician interprets the patient-text by accounting for him-it in terms of the explanatory network of science. Although the norms of the oral level will vary considerably, depending upon the norms of a given society and moral community, the norms of the literary level will be more constant since they depend more directly on the explanatory framework of science, which is less influenced by the type of values that divide moral communities.

Much of diagnosis will take place in these literary terms. But this will not be my focus in the remainder of this essay. Instead, I shall argue that diagnosis also serves as a link between oral and literary levels. It thus exemplifies a special form of the meeting of worlds; namely, an encounter that can be developed analogous to the activity of *writing*.

In diagnosis one has the personal relations of the oral level and the impersonal relations of the literary. Thus neither the model of dialogue nor that of interpretation is fully appropriate. But if one imagines a person dictating a letter to a scribe, then, in the act of writing the spoken words, we have an analogy with tremendous philosophical import.[26] It is a model that is central to several post-modern philosophies.

In diagnosis, the physician performs the role of an Egyptian scribe, translating the living spoken language of the patient into the written hieroglyph of the medical comunity. A process of transformation is used to divest the spoken word of all that is superfluous. The hieroglyph captures the essence and eliminates all else. In the same way, the physician strips away the "subjective" component in the patient's experience (including his self-description). The illness experience as a psychosomatic phenomenon is stripped of its psychic component. It becomes "disease". Metaphorically, the living patient is killed; divested of soul. And the physician is left with the corpse – the body and its functions. This process of killing the patient is the objectivization that is initiated by way of diagnosis.

One could also liken the activity to that of a Greek or Hebrew scribe. Let the psychic component be the vowel, the somatic the consonant. In the spoken word of illness the two are inseparable. Only in writing does the distinction appear. The scribe, in recording the event of the spoken word, takes the living unity and divides the sound, viewed as signifier, from the written word, viewed as linguistic signified. This distinction between spoken and written word appears *in* writing as the distinction between vowel and consonant. In ancient writing, only the consonants are recorded. In this way, the literary remnant (or trace) of the oral signifier is excluded. The scribe then takes the dead letter – the consonant – and buries it upon the page. This is the language of disease.

The time of the patient under the physician's care is like the time of the text, in the case of an epistle. During this time, the written word bewails its loss of dignity. But the time is necessary. Just as the scribe is, as it were, a necessary evil, necessitated by the distance between the speaker and hearer (in the case of an epistle); so the physician is a necessary evil, necessitated by the distance, which takes place in the objectivization of the body that accompanies the experience of pain. The goal of the patient in submitting to this mode of existence is not psychosomatic wholeness. The scribe's role is to restore integrity to the consonants. Then, hopefully, the breath of the vowel can be given to reanimate the dead letters.

Viewed in this way, the physician qua physician is not a friend.[27] He or she is a necessary evil – one that the patient would have preferred to avoid. The disease which the physician treats is the referent in the domain of writing that corresponds to the spoken word of illness. The world of disease – its grammar and syntax – is constructed in such a way that best facilitates the somatic healing that is the concern of medicine.[28] If one seeks to heal illness a different grammar and syntax is needed.[29]

All this does not involve the assumption that disease, as the somatic correlate of illness, is somehow a real, natural phenomenon, as opposed to a contructed phenomenon. Disease is itself a construct, just as body is a construct. Just as the distinction between vowel and consonant is the construct of a particular linguistic domain and varies among languages, so the mind-body distinction is a categorial construct that is culture and history dependent. Disease is a construct interdependent upon the mind-body construct. The nature of disease description involves a specification on the part of the medical community of the field of medical practice. This description is not fixed, but has a history. Maladies that today may be defined as psychological and thus outside the field of somatic medicine, may be redescribed in the future to lie within that medical domain. This redescription will be related to a redescription of the nature of mind and body, so that which was previously regarded as psychological will now be understood as somatic. The new paradigm will involve with it a new delineation of the appropriate field of medicine. This process of development will be moved by the tension between the ethical and technical ideals; by the tension between the tendency to narrow and to expand the scope of medicine. These tensions are, in turn, reflections of the difference between the oral and literary levels of the physician-patient relation and of the attempt of diagnosis to mediate the difference.

Given this model, the ethics of diagnosis can be developed as an ethic of writing. There is today a considerable literature that considers the ethical dimensions of orality and literacy. The analogy between diagnosis and writing thus opens up a dialogue between two fruitful fields. The degree to which this dialogue is mutually constructive must be further explored. But the following possibilities suggest themselves:

1. Writing, as an interrelation of oral and written levels, provides a model that may be used to develop the interrelation between the ethical and technical ideals of medicne. It provides a means for specifying and delineating the role-responsibility of the physician in such a way that concerns like objectivity and patient autonomy can be reconciled.

2. The interplay of signifier and signified is expressed at several levels and provides a way of developing the interrelation between: A. symptoms and disease; B. illness experience and disease description; C. clinic and laboratory; D. personal and impersonal dimensions of the physician-patient relation. By means of the sign, medical reality can be modeled.

3. The linguistic model provides a basis for a relation between the philosophy of medicine and the philosophy of language. It enables further development of the "literary metaphor" that has already been used by some physicians and philosophers of medicine.

Further discussion of these possibilities must await another time.

Institute of Religion
Houston, Texas

NOTES

* I would like to thank H. Tristram Engelhardt, Jr. for his critical comments and suggestions.

[1] For an account of the relation between mathematical, scientific reasoning and literacy see [17] and [18]. Havelock argues that the Newtonian revolution is to be closely associated with a convention that provides for calculation the same type of notation that alphabetic writing provides for speaking ([18], pp. 76–82, esp. p. 80). The connection between the rise of literacy and the decay of dialogue is well documented by Walter Ong [29]. Ong's general discussion of the decay of dialogue can shed much light on Reiser's account of the decay of dialogue in medicine [33].

[2] For the classical discussion of this see Augustine's *On Christian Doctrine* [2].

[3] For a good overview of Peirce's philosophy of language as well as the development in his thought on this theme, see [27]. In general one could say that Peirce's key contribution to modern linguistics rests on the observation that the copula in a proposition can be interpreted as a sign relation. Then all propositions can be regarded as instances of the signifier-signified relation. Logical relations can then be derived from fundamental linguistic ones. And logic in general can be developed as the science of signs. For Saussure's account of the "differential system of signs", as well as his formulation of the distinction between diachronic and synchronic approaches to language, see [37].

[4] In general, orality can be aligned with the pragmatic, and literacy with the essentialist. But this would be an alignment that is according to Ong's characterization and contrary to Derrida's. The issues are very complex but [4], [5], [12], [13], [17], [18], [29], and [30] give a representative sample.

[5] Although this approach will have relative merit, and it can be seen in the type of distinctions Amundsen makes in this volume (although Amundsen has a much greater appreciation of the interrelation between fact and value), the problems can be well illustrated by noting the difficulty we have in identifying the types of maladies that were diagnosed by a previous age ([7], p. 161). It is thus very difficult to isolate the technical from the ethical in order to evaluate a diagnosis by our standards of "technically good". And, as Amundsen notes, there will be several difficulties in evaluating the ethical merit of a diagnosis according to the ethical criteria of a past age. E.g. What does one say of a progressive physician whose practice is more "effective" but less in accord with the standards of his age?

[6] One could speak of it as "premodern" but not "pretechnical" in the sense of P. Lain-Entralgo' definition ([21], p. 14).

[7] In what follows, I shall continually refer to the account of Sydenham which is given by Miguel Sanchez-Gonzalez in his essay on medicine in John Locke's philosophy [35]. My reasons are twofold. First, Sanchez-Gonzalez makes all the main arguments that I wish to use in this essay, giving sufficient citations from the writings of Sydenham to document his case. But second, and more importantly, by showing the intimate relation between the medicine of Sydenham and the empirical philosophy of Locke, the particularities of medicine's development can be directly connected with the broader strokes that situate the development of the philosophy of language. This serves as a good point of departure for Foucault's brilliant account of how modern medicine's birth is not in some pure empiricism, as supposed, but in a restructuring of the relation between the visible and invisible, between signifier and signified [11]. The ideal of objectivity in the sense of Locke and

Sydenham, i.e. the ideal of the objective gaze as opposed to the objective content (ref. Descartes), while motivating the development of medicine (and philosophy), nevertheless proved untenable. This objective gaze, however, is the stare that kills the patient and leaves a corpse that can be explored. It then gives the condition of the particular outer/inner relation that becomes the dialectic of clinic and laboratory. For a more comprehensive account of Sydenham and Locke, see [34].

8 [35], p. 680. The word "essence" is ambiguous, however. In another sense, one could say that Sydenham seeks to know the essence of a disease apart from knowing its cause and its underlying mechanisms (like one can know the genus and species of a plant without knowing its underlying mechanisms). See [40], pp. 146–147 for the genus/species analogy. [11], p. 7f. develops some implications of this "botanical model".

9 "Medicine doubtlessly should be considered as the prototype of this type of observational sciences which though incapable of offering certainty, can provide man the advantages of comfort and health" ([35], p. 681).

10 See the essays in this volume, esp. [10].

11 For differing accounts of this restructuring see [7], pp. 176 –184; [32], [11].

12 In a more comprehensive discussion, we would have to consider three ways in which observer bias is checked: (1) the laboratory, (2) statistics, (3) controlled experiment. But the second two can be regarded as ways of controlling and confirming the integration of the pragmatic and essentialist criteria. They check noise, while the laboratory puts a check on the theory-infectedness of knowledge. (See note 15 on the meanings of "observer bias".)

13 The account which Leon Kass gives of the end of medicine would provide a good basis for this alternative interpretation [20]. The idealogy that goes with the dialetic of clinic and laboratory can be viewed as a way of restricting medicine to somatic concerns. Later in this essay, I shall argue that an account something like that advanced by Kass should serve as a regulative ideal for medicine. But it is problematic if it is viewed as descriptive of present practice.

14 [3], pp. 57–58. The story is from [44], pp. 117–130. I use the story to illustrate a different point than Carson's .

15 "Observer bias" can mean two very different things. On one hand, it can refer to the *noise* that hinders precise recording of the data of observation. Then it is a clearly negative phenomenon, and one seeks to overcome it by means such as statistical analysis and controlled experiment. On the other hand, "observer bias" can be used to designate the subjective, *theory-infected component* of any observation. This component is not necessarily negative. It may pertain to the general conditions of all human knowing (ref. Kant's account of the conditions of human knowledge). Then it is not necessarily contrary to objectivity if one takes 'objective' in Kant's reconstructed sense. "Observer bias" could also refer to the conditions that qualify the unique perspective of a particular individual. Then it refers to "subjectivity". In what follows, I shall use 'observer bias' to designate the conditionedness of human knowledge rather than noise.

16 Consider, for example, Pellegrino's attempt to incorporate value desiderate into the diagnostic process [31]. The computer takes over the more rigorously scientific factors and the physician works with the computer by inputing what may be viewed as the patient's observer bias. (Here I use 'observer bias' in the sense of "subjective component", see note 15). Sass also shows how values can be positively incorporated into the diagnostic process [36].

17 Here one takes a Kantian-type approach, developing the conditions of the possibility of

knowledge, but the established conditions are regulative (like Kant's ideal of pure reason) rather than constitutive. The approach is not dissimilar to that of Peirce. [8], esp. p. 8 develops the implications of this for controversies in medicine.

[18] Amundsen suggests this when he notes that "when one's physician exhausts pathoanatomical and pathophysiological avenues of diagnosis and pursues a pathopsychological mode of inquiry, the patient may well become uncomfortable and feel that the physician is acting unethically" ([1], p. 59).

[19] [20], p. 10. If one seeks a broader concept of medicine, one could say that the patient gives the 'object body' to science. (For a good overview of the meaning of 'object body' see [14]. For a more comprehensive discussion, see [38]). In this essay, I will leave unresolved the question as to whether the physiological body or object body should be the focus of medicine. The important point for this essay is that the narrowing of medicine to body can be sustained in both a cartesian and a 'lived body' approach to the mind-body relation.

[20] Mainetti well notes how the gap that opens between the self one is and the self one has leads to the transcendence called culture ([26], p. 85). In this case, the culture is the institution of medicine and the paradox is that objectivization of the patient (brought about by pain) is overcome by the development of a medicine that increasingly objectivizes the patient.

[21] Such distance is central to the meaning of objectivity. It is also of central importance to the foundation of Hippocratic medicine [21], and to the development and advancement of modern medicine (see e.g. Reiser's discussion of the stethoscope in[32] and [33]).

[22] Note, for example, the way a teenage girl's broader problem is resolved, once the physician heals her complexion (the fictional story is summarized in [3], pp. 57–58).

[23] This is one way to respond to Spicker's account of Dewitt Stetton ([39], pp. 119). One could then rephrase Spicker's comment to read: "The failure of contemporary physicians to earn the honor of 'good person' [where 'good person' replaces 'good physician'] can be found in their attitudes toward handicapped (or partially handicapped) patients". The difficulty with even this interpretation is that 'good' in 'good physician' can be interpreted ethically and technically, and there is both an overlap and a tension between the two.

[24] A good example of this can be found in H.-G. Gadamer's discussion of the 'Zeitenabstand' ([13], p. 63). Just as the temporal distance is not something to be overcome, but rather should be appropriated as a productive possibility of understanding, so the distance initiated by e.g. the stethoscope should be appropriated as a productive possibility of medical intervention.

[25] This is well expressed by Wartofsky and Zaner when they note that "the relation between physician and patient is never simply that of an 'I' to an 'it', but neither is it only that of an 'I' to a 'thou'" ([43], p. 3). Both simultaneously characterize the relation.

[26] Note the similarity between this analogy and the psychoanalytic one that Reiser uses ([32], [33]).

[27] This goes well with Patricia Illingworth's criticism of the friendship model of the physician-patient relation [19], although her discussion of patient autonomy would need to be further qualified.

[28] One could say that it is the syntax and grammar of *explanation* rather than understanding (see [43], [25]).

[29] Then one needs the syntax and grammar of *understanding*. One could also view the grammar and syntax of disease (and of explanation in general) as a limiting case of illness (and of understanding), much like Newtonian mechanics can be viewed as a limiting case of special relativity.

BIBLIOGRAPHY

1. Amundsen, D.W.: 1992, 'Some Conceptual and Metholodological Observations on the History of Ethics of Diagnosis', in this volume, pp. 47–61.
2. Augustine: 1983, *On Christian Doctrine*, J.F. Shaw (tr.), in *Nicene and Post-Nicene Fathers* Vol.**II**, R. Schaff (ed.), Wm.B Eardmans Publishing Company, Grand Rapids, Michigan.
3. Carson, R.A.: 1990, 'Interpretive Bioethics: The Way of Discernment', *Theoretical Medicine* **11**, 51–59.
4. Derrida, J.: 1974, *Of Grammatology*, G.C. Spivak (tr), Johns Hopkins University Press, Baltimore, Maryland.
5. Derrida, J.: 1978, *Writing and Dfference*, A. Bass (tr.), University of Chicago Press, Chicago.
6. Engelhardt, H. T.: 1982, 'Goals of Medical Care: A Reappraisal', in N.K. Bell (ed.), *Who Decides: Conflicts of Rights in Health Care*, The Humana Press, New Jersey.
7. Engelhardt, H.T.: 1986, *The Foundations of Bioethics*, Oxford University Press, New York/Oxford.
8. Engelhardt, H.T. and Caplan, A.L.: 1987, 'Patterns of Controversy and Closure: The Interplay of Knowledge, Values, and Political Forces', in H.T. Engelhardt and A.L. Caplan (eds.), *Scientific Controversies*, Cambridge University Press, Cambridge, pp. 1–23.
9. Engelhardt, H.T.: 1991, *Bioethics and Secular Humanism: The Search for a Common Morality*, SCM Press, London and Trinity Press International, Philadelphia.
10. Engelhardt, H.T.: 1992, 'Observer Bias: The Emergence of the Ethics of Diagnosis', in this volume, pp. 63–71.
11. Foucault, M.: 1975, *The Birth of the Clinic*, A.M. Smith (trans.), Vintage Books, New York.
12. Gadamer, H.-G.: 1986, *Wahrheit und Methode: Grundzüge einer philosophischen Hermeneutik*, J.C.B. Mohr, Tübingen.
13. Gadamer, H.-G.: 1986, *Wahrheit und Methode: Ergänzungen Register*, J.C.B. Mohr, Tübingen.
14. Gadow, S.: 1980, 'Body and self: A Dialectic', *Journal of Medicine and Philosophy* **5**(3), 172–185.
15. Gogel, E.L. and Terry, J.S.: 1987, 'Medicine as Interpretation: The Uses of Literary Metaphors and Methods', *Journal of Medicine and Philosophy* **12**, 205–217.
16. Harding, S.: 1978, 'Knowledge, Technology, and Social Relations', *Journal of Medicine and Philosophy* **3**(4), 346–358.
17. Havelock, E.A.: 1967, *Preface to Plato*, Grosset and Dunlap by arrangement with the Harvard University Press, New York.
18. Havelock, E.A.: 1976, *Origins of Western Literacy*, Ontario Institute for Studies in Education, Toronto.
19. Illingworth, P.M.L.: 1988, 'The Friendship Model of Physician/Patient Relationship and Patient Autonomy', *Bioethics* **2**(1), 22–36.
20. Kass, L.R.: 1981, 'Regarding the End of Medicine and the Pursuit of Health', in A.L. Caplan *et al.*, *Concepts of Health and Disease*, Addison-Wesley Publishing Co., London, pp. 3–30.
21. Laín-Entralgo, p.: 1992, 'The Ethics of Diagnosis in Ancient Greek Medicine', in this volume, pp. 13–18.

22. Leder, D.: 1990, 'Clinical Interpretation: The Hermeneutics of Medicine', *Theoretical Medicine* **11**, 9–24.
23. Leder, D.: 1992, 'The Experience of Pain and its Clinical Implications', in this volume, pp. 95–105.
24. Levinas, E.: 1969, *Totality and Infinity*, A. Lingis (tr.), Duquesne University Press, Pittsburgh.
25. Lock, J.D.: 1990, 'Some Aspects of Medical Hermeneutics: The Role of Dialectic and Narrative', *Theoretical Medicine* **11**, 41–49.
26. Mainetti, J.A.: 1992 'Embodiment, Pathology and Diagnosis', in this volume, pp. 79–93.
27. Murphey, M.G.: 1961, *The Development of Peirce's Philosophy*, Cambridge, Mass.
28. Murphy, E.A: 1978, 'Some Epitemological Aspects of the Model in Medicine', *Journal of Medicine and Philosophy* **3**(4), 273–292.
29. Ong, W.J.: 1958, *Ramus: Method, and the Decay of Dailogue*, Harvard University Press, Cambridge, Mass.
30. Ong, W.J.: 1982, *Orality and Literacy: The Technologizing of the Word*, Methuen & Co., Ltd., London.
31. Pellegrino, E.D.: 1992 'Value Desiderata in the Logical Structuring of Computer Diagnosis', in this volume, pp. 173–195.
32. Reiser, S.J.: 1978, *Medicine and the Reign of Technology*, Cambridge University Press, New York.
33. Reiser, S.J.: 1978, 'The Decline of the Clinical Dialogue', *Journal of Medicine and Philosophy* **3**(4), 305–313.
34. Romanell, P.: 1987, *John Locke and Medicine*, Prometheus Books, New York.
35. Sanchez-Gonzalez, M.A.: 1990, 'Medicine in John Locke's Philosophy', *Journal of Medicine and Philosophy* **15**, 6775–695.
36. Sass, H.-M.: 1992, 'Diagnosing the Eleven Month Pregnancy', in this volume, pp. 153–162.
37. Saussure, F.: 1966, *Course in General Linguistics*, W. Baskin (trans.), McGraw-Hill Book Company, New York.
38. Spicker, S. (ed.): 1970, *The Philosophy of the Body: Rejections of Cartesian Dualism*, Quadrangle/New York Times Publishing Co., New York.
39. Spicker, S.: 1992, 'Ethics in Diagnosis: Bodily Integrity, Trust-Telling, and the Good Physician', in this volume, pp. 107–122.
40. Sydenham, T.: 1981, 'Preface to the Third Edition, *Observations Medicae*' in A. Caplan, *et al.* (eds.), *Concepts of Health and Disease*, Addison-Wesley Publishing Co., London, pp. 145–156.
41. Turner, B.: 1987, *Medical Power and Social Knowledge*, Sage Publications, London.
42. Wartofsky, M.W.: 1978, 'Editorial', *Journal of Medicine and Philosophy* **3**(4), 265–272.
43. Wartofsky, M.W. and Zaner, R.M.: 1980, 'Editorial', Journal of *Medicine and Philosophy* **5**(1), 1–7.
44. Williams, W.C.: 1961, *The Farmers Daughters*, New Directives, New York.

NOTES ON CONTRIBUTORS

Augustín Albarracín , M.D., is Professor, Centro de Estudios Historicos of the Consejo Superior de Investigacions Cientificas, and Professor, Department of History of Medicine in the Faculty of Medicine of the Autonomous University of Madrid, Madrid, Spain.

Darrel Amundsen, Ph.D., is Professor, Department of Foreign Language and Literature, Western Washington State University, Bellingham, Washington 98225, U.S.A.

Eugene V. Boisaubin, M.D., is Associate Professor, Department of Medicine, and Member, Center for Ethics, Medicine, and Public Issues, Baylor College of Medicine, Houston, Texas 77030, U.S.A.

Thomas J. Bole, III, Ph.D., is Program Associate, Biomedical and Health Care Ethics Program, University of Oklahoma Health Sciences Center, Oklahoma City, Oklahoma 73190, U.S.A

Mary Ann Gardell Cutter, Ph.D., is Assistant Professor, University of Colorado, Colorado Springs, Colorado.

H. Tristram Engelhardt, Jr., is Professor, Departments of Medicine as well as Community Medicine and Obstetrics and Gynaecology, Baylor College of Medicine, Houston, Texas 77030, U.S.A. In addition, he is Professor in the Department of Philosophy, Rice University, Adjunct Research Fellow, Institute of Religion, and Member of the Center for Ethics, Medicine, and Public Issues.

Diego Gracia, M.D., is Professor, Department of History of Medicine in the Faculty of Medicine of the Complutensis University of Madrid, Madrid, Spain.

George Khushf, M.A., is Research Associate, Institute of Religion, Houston, Texas 77030, U.S.A.

Pedro Laín-Entralgo, Ph.D., is Professor, Department of History of Medicine in the Faculty of Medicine of the Complutensis University of Madrid, Madrid, Spain.

Drew Leder, M.D., Ph.D., is Assistant Professor, Department of Philosophy, Loyola College in Maryland, Baltimore, Maryland 21220, U.S.A.

José Alberto Mainetti, Ph.D., M.D., Professor, Chair of Medical Humanities, Faculty of Medical Sciences, La Plata National University, Argentina.

Edmond A. Murphy, M.D., is Professor, Division of Medical Genetics, Johns Hopkins University, School of Medicine, Baltimore, Maryland 21286, U.S.A.

Edmund D. Pellegrino, M.D., is Director, Center for the Advanced Study of Ethics, and Director, Center for Clinical Bioethics, Georgetown University, Washington, D.C. 20057, U.S.A.

José Luis Peset, M.D., is Professor, Centro de Estudios Historicos of the Consejo Superior de Investigaciones Cientificas, Madrid, Spain.

Hans-Martin Sass, Ph.D., is Professor, Institut für Philosophie, Ruhr-Universität Bochum, and Georgetown University, the Kennedy Institute of Ethics and the Medical School, Washington, D.C. 20057, U.S.A.

Kenneth F. Schaffner, M.D., Ph.D., is University Professor of Medical Humanities, Room

J. L. Peset and D. Gracia (eds.), The Ethics of Diagnosis, 301–302.

714T Gelman Library, The George Washington University, Washington, DC 20052, U.S.A.

Stuart F. Spicker, Ph.D., Professor, Division of Humanistic Studies in Medicine, Department of Community Medicine and Health Care, School of Medicine, University of Connecticut Health Center, Farmington, Connecticut 06032, U.S.A.

Marx W. Wartofsky, Ph.D., is Professor, Department of Philosophy, The City University of New York, Baruch College and Graduate Center, New York, New York 10010, U.S.A.

Henrik R. Wulff, M.D., is Professor in Clinical Theory and Ethics, Unit of Medical Philosophy and Clinical Theory, Panum Institute, University of Copenhagen, 2200 Copenhagen N, Denmark.

INDEX

Philosophy and Medicine

1. H. Tristram Engelhardt, Jr. and S.F. Spicker (eds.): *Evaluation and Explanation in the Biomedical Sciences.* 1975   ISBN 90-277-0553-4
2. S.F. Spicker and H. Tristram Engelhardt, Jr. (eds.): *Philosophical Dimensions of the Neuro-Medical Sciences.* 1976   ISBN 90-277-0672-7
3. S.F. Spicker and H. Tristram Engelhardt, Jr. (eds.): *Philosophical Medical Ethics: Its Nature and Significance.* 1977   ISBN 90-277-0772-3
4. H. Tristram Engelhardt, Jr. and S.F. Spicker (eds.): *Mental Health: Philosophical Perspectives.* 1978   ISBN 90-277-0828-2
5. B.A. Brody and H. Tristram Engelhardt, Jr. (eds.): *Mental Illness.* Law and Public Policy. 1980   ISBN 90-277-1057-0
6. H. Tristram Engelhardt, Jr., S.F. Spicker and B. Towers (eds.): *Clinical Judgment: A Critical Appraisal.* 1979   ISBN 90-277-0952-1
7. S.F. Spicker (ed.): *Organism, Medicine, and Metaphysics.* Essays in Honor of Hans Jonas on His 75th Birthday. 1978   ISBN 90-277-0823-1
8. E.E. Shelp (ed.): *Justice and Health Care.* 1981
                    ISBN 90-277-1207-7; Pb 90-277-1251-4
9. S.F. Spicker, J.M. Healey, Jr. and H. Tristram Engelhardt, Jr. (eds.): *The Law-Medicine Relation: A Philosophical Exploration.* 1981   ISBN 90-277-1217-4
10. W.B. Bondeson, H. Tristram Engelhardt, Jr., S.F. Spicker and J.M. White, Jr. (eds.): *New Knowledge in the Biomedical Sciences.* Some Moral Implications of Its Acquisition, Possession, and Use. 1982   ISBN 90-277-1319-7
11. E.E. Shelp (ed.): *Beneficence and Health Care.* 1982   ISBN 90-277-1377-4
12. G.J. Agich (ed.): *Responsibility in Health Care.* 1982   ISBN 90-277-1417-7
13. W.B. Bondeson, H. Tristram Engelhardt, Jr., S.F. Spicker and D.H. Winship: *Abortion and the Status of the Fetus.* 2nd printing, 1984   ISBN 90-277-1493-2
14. E.E. Shelp (ed.): *The Clinical Encounter.* The Moral Fabric of the Patient-Physician Relationship. 1983   ISBN 90-277-1593-9
15. L. Kopelman and J.C. Moskop (eds.): *Ethics and Mental Retardation.* 1984
                    ISBN 90-277-1630-7
16. L. Nordenfelt and B.I.B. Lindahl (eds.): *Health, Disease, and Causal Explanations in Medicine.* 1984   ISBN 90-277-1660-9
17. E.E. Shelp (ed.): *Virtue and Medicine.* Explorations in the Character of Medicine. 1985   ISBN 90-277-1808-3
18. P. Carrick: *Medical Ethics in Antiquity.* Philosophical Perspectives on Abortion and Euthanasia. 1985   ISBN 90-277-1825-3; Pb 90-277-1915-2
19. J.C. Moskop and L. Kopelman (eds.): *Ethics and Critical Care Medicine.* 1985
                    ISBN 90-277-1820-2
20. E.E. Shelp (ed.): *Theology and Bioethics.* Exploring the Foundations and Frontiers. 1985   ISBN 90-277-1857-1
21. G.J. Agich and C.E. Begley (eds.): *The Price of Health.* 1986
                    ISBN 90-277-2285-4
22. E.E. Shelp (ed.): *Sexuality and Medicine.*
    Vol. I: Conceptual Roots. 1987   ISBN 90-277-2290-0; Pb 90-277-2386-9

23. E.E. Shelp (ed.): *Sexuality and Medicine.*
Vol. II: Ethical Viewpoints in Transition. 1987
ISBN 1-55608-013-1; Pb 1-55608-016-6
24. R.C. McMillan, H. Tristram Engelhardt, Jr., and S.F. Spicker (eds.): *Euthanasia and the Newborn.* Conflicts Regarding Saving Lives. 1987
ISBN 90-277-2299-4; Pb 1-55608-039-5
25. S.F. Spicker, S.R. Ingman and I.R. Lawson (eds.): *Ethical Dimensions of Geriatric Care.* Value Conflicts for the 21th Century. 1987
ISBN 1-55608-027-1
26. L. Nordenfelt: *On the Nature of Health.* An Action- Theoretic Approach. 1987
ISBN 1-55608-032-8
27. S.F. Spicker, W.B. Bondeson and H. Tristram Engelhardt, Jr. (eds.): *The Contraceptive Ethos.* Reproductive Rights and Responsibilities. 1987
ISBN 1-55608-035-2
28. S.F. Spicker, I. Alon, A. de Vries and H. Tristram Engelhardt, Jr. (eds.): *The Use of Human Beings in Research.* With Special Reference to Clinical Trials. 1988
ISBN 1-55608-043-3
29. N.M.P. King, L.R. Churchill and A.W. Cross (eds.): *The Physician as Captain of the Ship.* A Critical Reappraisal. 1988
ISBN 1-55608-044-1
30. H.-M. Sass and R.U. Massey (eds.): *Health Care Systems.* Moral Conflicts in European and American Public Policy. 1988
ISBN 1-55608-045-X
31. R.M. Zaner (ed.): *Death: Beyond Whole-Brain Criteria.* 1988
ISBN 1-55608-053-0
32. B.A. Brody (ed.): *Moral Theory and Moral Judgments in Medical Ethics.* 1988
ISBN 1-55608-060-3
33. L.M. Kopelman and J.C. Moskop (eds.): *Children and Health Care.* Moral and Social Issues. 1989
ISBN 1-55608-078-6
34. E.D. Pellegrino, J.P. Langan and J. Collins Harvey (eds.): *Catholic Perspectives on Medical Morals.* Foundational Issues. 1989
ISBN 1-55608-083-2
35. B.A. Brody (ed.): *Suicide and Euthanasia.* Historical and Contemporary Themes. 1989
ISBN 0-7923-0106-4
36. H.A.M.J. ten Have, G.K. Kimsma and S.F. Spicker (eds.): *The Growth of Medical Knowledge.* 1990
ISBN 0-7923-0736-4
37. I. Löwy (ed.): *The Polish School of Philosophy of Medicine.* From Tytus Chałubiński (1820–1889) to Ludwik Fleck (1896–1961). 1990
ISBN 0-7923-0958-8
38. T.J. Bole III and W.B. Bondeson: *Rights to Health Care.* 1991
ISBN 0-7923-1137-X
39. M.A.G. Cutter and E.E. Shelp (eds.): *Competency.* A Study of Informal Competency Determinations in Primary Care. 1991
ISBN 0-7923-1304-6
40. J.L. Peset and D. Gracia (eds.): *The Ethics of Diagnosis.* 1992
ISBN 0-7923-1544-8
41. K.W. Wildes, S.J., F. Abel, S.J. and J.C. Harvey (eds.): *Birth, Suffering, and Death.* Catholic Perspectives at the Edges of Life. 1992
ISBN 0-7923-1547-2

Philosophy and Medicine

KLUWER ACADEMIC PUBLISHERS – DORDRECHT / BOSTON / LONDON